Isabel Urbat

Oceanic anoxic event (OAE) 1b

Isabel Urbat

Oceanic anoxic event (OAE) 1b

High resolution geochemical studies,
Mazagan Plateau and the Vocontian Basin

Südwestdeutscher Verlag für Hochschulschriften

Impressum/Imprint (nur für Deutschland/ only for Germany)
Bibliografische Information der Deutschen Nationalbibliothek: Die Deutsche Nationalbibliothek verzeichnet diese Publikation in der Deutschen Nationalbibliografie; detaillierte bibliografische Daten sind im Internet über http://dnb.d-nb.de abrufbar.
Alle in diesem Buch genannten Marken und Produktnamen unterliegen warenzeichen-, marken- oder patentrechtlichem Schutz bzw. sind Warenzeichen oder eingetragene Warenzeichen der jeweiligen Inhaber. Die Wiedergabe von Marken, Produktnamen, Gebrauchsnamen, Handelsnamen, Warenbezeichnungen u.s.w. in diesem Werk berechtigt auch ohne besondere Kennzeichnung nicht zu der Annahme, dass solche Namen im Sinne der Warenzeichen- und Markenschutzgesetzgebung als frei zu betrachten wären und daher von jedermann benutzt werden dürften.

Verlag: Südwestdeutscher Verlag für Hochschulschriften Aktiengesellschaft & Co. KG
Dudweiler Landstr. 99, 66123 Saarbrücken, Deutschland
Telefon +49 681 37 20 271-1, Telefax +49 681 37 20 271-0
Email: info@svh-verlag.de
Zugl.: Köln, University, Diss., 2009

Herstellung in Deutschland:
Schaltungsdienst Lange o.H.G., Berlin
Books on Demand GmbH, Norderstedt
Reha GmbH, Saarbrücken
Amazon Distribution GmbH, Leipzig
ISBN: 978-3-8381-1723-2

Imprint (only for USA, GB)
Bibliographic information published by the Deutsche Nationalbibliothek: The Deutsche Nationalbibliothek lists this publication in the Deutsche Nationalbibliografie; detailed bibliographic data are available in the Internet at http://dnb.d-nb.de.
Any brand names and product names mentioned in this book are subject to trademark, brand or patent protection and are trademarks or registered trademarks of their respective holders. The use of brand names, product names, common names, trade names, product descriptions etc. even without a particular marking in this works is in no way to be construed to mean that such names may be regarded as unrestricted in respect of trademark and brand protection legislation and could thus be used by anyone.

Publisher: Südwestdeutscher Verlag für Hochschulschriften Aktiengesellschaft & Co. KG
Dudweiler Landstr. 99, 66123 Saarbrücken, Germany
Phone +49 681 37 20 271-1, Fax +49 681 37 20 271-0
Email: info@svh-verlag.de

Printed in the U.S.A.
Printed in the U.K. by (see last page)
ISBN: 978-3-8381-1723-2

Copyright © 2010 by the author and Südwestdeutscher Verlag für Hochschulschriften Aktiengesellschaft & Co. KG and licensors
All rights reserved. Saarbrücken 2010

Zusammenfassung

Während der Kreide wurden zahlreiche Störungen des globalen Kohlenstoffkreislaufes aufgezeichnet. Diese Störungen führten zu organisch reichen Ablagerungen, den sogenannten ozeanischen anoxischen Ereignissen (OAEs). Der OAE 1b, abgelagert in der Unteren Kreide, ist einer dieser bedeutenden CO_2-Störungen und wird mit der Sedimentation von Schwarzschiefern in Verbindung gebracht. Beschrieben wurde OAE 1b in Ablagerungen des Unteren Albs für die Regionen des Nord- und Zentral Atlantiks sowie im westlichen Einzugsbereich der Tethys und wird deshalb als überregionaler Schwarzschiefer beschrieben. OAE 1b ist assoziiert mit einer abrupten Verschiebung zu ausgeprägt negativen Werten innerhalb zahlreicher Kohlenstoffisotopenreservoirs. Negative Werte in den Kohlenstoffisotopen wurden ebenfalls für das OAE im Toarcium und das OAE 1a im Frühen Apt beschrieben und werden mit spontaner Freisetzung von Methan aus marinen Gashydraten in Zusammenhang gestellt. In dieser hier präsentierten Studie wurden hochauflösende geochemische Profile von zwei paläogeographisch getrennten und klimatisch unterschiedlichen Gebieten (Mazagan Plateau DSDP Site 545 und dem Vokontischen Becken) erstellt. Während dem Frühen Alb befand sich das Mazagan Plateau innerhalb des nördlichen tropischen Klimagürtel vor der Küste Marokko und das Vokontische Becken befand sich im subtropischen Klimagürtel innerhalb des Areals der westlichen Tethys. Ergebnisse bestätigen, dass das Schwarzschiefer Ereignis durch die $\delta^{13}C$ Isotope des Gesamt- karbonatischen und -organischen Kohlenstoffs definiert werden kann. Der Beginn und das Ende der Sedimentation sind zeitgleich mit den negativen Isotopen Werten für beide Untersuchungsgebiete. Komponentenspezifische Kohlenstoff Isotope, gemessen für das tropische Gebiet Site 545, zeigen darüber hinaus, dass die Störung des Kohlenstoffkreislaufes zeitgleich in marinen und atmosphärischen Kohlenstoffreservoirs stattfand. Dies ist angedeutet durch die Zusammensetzung stabiler Biomarkerisotope für terrestrische höhere Pflanzenwachse sowie von marinen Algenlipiden.

Auffallend ist ein plötzlicher Anstieg der Meeresoberflächentemperatur um 3 °C, gemessen an der tropisch gelegenen Site 545 mit dem TEX_{86} Temperaturproxy, zeitgleich mit den negativen Werten in den Kohlenstoffisotopen zu Beginn des OAE 1b. Ein Temperaturanstieg mit einsetzender OAE 1b Sedimentation ist ebenso für das Vokontische Becken angedeutet und lässt auf eine schnelle, globale Erwärmung für

ZUSAMMENFASSUNG

diesen Zeitabschnitt schließen. An beiden untersuchten Gebieten deuten Klimaproxys, veranschaulicht durch Element Verhältnisse auf Al normalisiert, auf veränderte Klimakonditionen zu Beginn und während des OAE 1b hin. Am Mazagan Plateau zeigen Schwankungen innerhalb der Zusammensetzung des terrestrischen Materials zyklische warm/kalt Variationen an, die von Küstenauftriebsströmungen gesteuert werden. Der detritische Eintrag (Si/Al) besteht überwiegend aus äolischem Staub, welcher über die NE-Passatwinde vom Afrikanischen Kontinent aus, eingetragen wurde. Mit der einsetzenden OAE 1b Sedimentation verlangsamte sich der Küstenauftrieb, angedeutet durch verlangsamte Wind Systeme, und humideres Klima etablierte sich. Das Klimasystem spiegelte bis kurz nach Abschluss der Schwarzschiefersedimentation diesen gestörten Zustand wider. Zyklische Schwankungen stellen sich erst im oberen Bereich des Datenprofils wieder ein.

Das Vokontische Becken spiegelt ein eingeschränkteres flachmarines Milieu wider, das zeitlich begrenzt von Karbonatplattformen innerhalb der Tethys umgeben war. Der allgemeine Klimatrend ist von orbitalen Variationen gesteuert, im wesentlichen der Exzentrizität. Vor dem OAE 1b ist eine geringere Amplitude in zyklischen Schwankungen des Präzessionssignals vorherrschend, dokumentiert in der siliziklastischen Fraktion. Während des OEAs deuten verstärkte Amplituden der Zyklik auf verstärkte Klimakontraste hin resultierend in abwechselnd hoher Aridität und verstärkter Humidität. Nach dem Schwarzschiefer Ereignis bleibt das System für mehrere tausend Jahre gestört bevor sich prä-OAE Klimaverhältnisse wieder einstellen.

Die Auswirkungen dieser massiven Störung des Klimas hatte ebenfalls Folgen für die Flora und Fauna an beiden untersuchten Gebieten. Am Mazagan Plateau nahm die Produktion von organischem Material zu, obwohl die Küstenauftriebsströmung, welche mehr als zweihunderttausend Jahre lang vor dem Event vorherrschend war, schwächer wurde. Erhöhte Produktivität resultierte vermutlich aus erhöhtem Nährstoffeintrag, geliefert vom angrenzenden afrikanischen Kontinent. Die Menge der Bioproduktion veränderte sich nicht für das Vokontische Becken. Die Zusammensetzung aber verschob sich hin zu bakteriendominierten Lebensgemeinschaften, angezeigt von ausgewählten Isoprenoiden (z. B. Lycopan). Erhöhte Erhaltungsbedingungen begünstigten die Anreicherung von organischem Material. Die gemessenen Daten dokumentieren eine Erwärmungsphase während dem OAE 1b, die zu abgeschwächten Passatwinden innerhalb der Tropen und stärkeren Klimakontrasten in den Subtropen führte.

ZUSAMMENFASSUNG

Ein weiterer wichtiger Bestandteil dieser Arbeit ist die Ergründung des Auslöseimpulses, der zu dem abrupten Einsetzen und Abschluss der OAE 1b Sedimentation führte. Die Magnitude des Isotopenexkurses am Mazagan Plateau bleibt über mehrere 10000 Jahre lang konstant. Im Gegensatz dazu würde die Freisetzung eines einzelnen Methanereignisses das Atmosphärensignal zwar stark beeinträchtigen, leichte Kohlendioxidkonzentrationen würden allerdings gleichmäßig und fortwährend abgebaut. Die Daten vom Mazagan Plateau sprechen hingegen für eine konstante Zufuhr von ^{13}C-abgereicherten CO_2 während des gesamten Schwarzschiefer Intervalls. Es wurde deshalb eine Kombination, bestehend aus einzelnen, moderaten Methanereignissen innerhalb kontinuierlich erhöhter Methanemissionen über einen längeren Zeitraum hinweg, angenommen. Die zwei Ereignisse grenzen zeitlich den Auslöseimpuls des OAEs auf etwa 25000 Jahre ein. Berechnet mit einem gekoppelten Atmosphären-Land-Ozean-Sediment Modell für das Mazagan Plateau weisen die Ergebnisse darüber hinaus Ähnlichkeiten zum Datensatz des Vokontischen Beckens auf. Beide Klimasysteme reagierten trotz Ähnlichkeiten sehr unterschiedlich auf die Störungsphase, welche einherging mit dem Beginn der OAE Sedimentation. Diese Unterschiede können durch eine Verschiebung der Inter Tropischen Konvergenz Zone (ITCZ) erklärt werden. Mithilfe eines Atmospheric General Circulation Model (AGCM) wurden netto Fluxraten kalkuliert. Während der Kreide zeigen berechnete Niederschlags- und Verdunstungsraten für die untersuchten Bereiche, starke regionale Unterschiede in Abhängigkeit von den Paläobreitengraden an. Eine moderate Verschiebung der ITCZ um ~15°N während des OAE 1b, erklärt die einsetzende Humidität verbunden mit abflachenden Passatwinden am Mazagan Plateau und ebenfalls die verstärkten Klimakontraste, mit allgemeinen Trend zu arideren Verhältnisse am Vokontischen Becken. Der OAE 1b ist an zahlreichen verschiedenen Lokationen beschrieben worden. Alle weisen spezielle klimatische und ökologische Besonderheiten auf. Aufgrund der charakteristischen C-Isotopen Kurven und Temperatur Proxys sind diese trotzdem vergleichbar. Des weiteren weisen die zeitgleiche und weitverbreitete Ablagerung sowie Vergleiche mit der Literatur, den OAE 1b eher als ein globales anstelle eines überregionalen Ereignisses aus.

Abstract

The Cretaceous is characterised by several perturbations of the global carbon cycle, resulting in the so–called Oceanic Anoxic Events (OAEs). During the Lower Cretaceous the OAE 1b is one of these major perturbation associated with black shale deposition. It is documented in Lower Albian deposits from the North and Central Atlantic region as well as from the western Tethys area and is therefore considered a supra-regional black shale. OAE 1b is associated with a pronounced negative shift in various carbon isotope reservoirs. Negative shifts in the carbon isotopes are also observed for the Toarcian OAE and the early Aptian OAE 1a and are thought to result from the catastrophic release of methane from marine gas hydrates. In the here presented study, high-resolution geochemical profiles from two paleogeographically distinct locations (Mazagan Plateau DSDP Site 545 and the Vocontian Basin) are established. The Mazagan Plateau was located in the lower Albian northern tropical climate belt offshore Morocco and the Vocontian Basin was situated in the subtropical Albian climate belt in the western Tethys region. The results show, that the black shale interval is defined by the bulk carbonate and bulk organic carbon $\delta^{13}C$ isotope records, because it starts and ends simultaneously with the negative isotope excursion at both sites. Compound-specific carbon isotopes for the tropical site 545 show furthermore, that the perturbation of the carbon cycle occurred synchronously in the oceanic- and atmospheric carbon reservoirs. This is indicated by the stable isotopic composition of biomarkers for terrestrial higher plant waxes and marine algal lipids.

Strikingly, an abrupt 3 C° rise in sea surface temperature (SST), documented by the TEX_{86} temperature proxy, at the tropical site 545 is observed parallel to the negative shift in the carbon isotopes at the onset of OAE 1b. A temperature increase is also indicated for the onset of the OAE 1b interval in the Vocontian Basin, suggesting rapid global warming in that time period. For both investigated sites, climate proxies based on Al-normalized elemental ratios indicate changing climate conditions at the onset and during the black shale deposition. Prior to OAE 1b, fluctuations in the terrestrial composition at the Mazagan Plateau indicate cyclic warm-cold variations, related to upwelling conditions. The detrital input (Si/Al) is dominated by the eolian dust, delivered by the NE-trade wind system from the African continent. With the onset of OAE 1b the upwelling cell slows down, as is indicated by the decreasing ratios of the wind derived proxies, and more humid

Abstract

climate conditions develop. The climate system remains disturbed until shortly after the black shale termination. Cyclic fluctuations re-establish at the upper part of the record.

The Vocontian Basin was a more restricted shallow marine environment, temporally surrounded by carbonate platforms within the Tethys realm. The general climate trend is influenced by eccentricity driven orbital variations. Prior to OAE 1b low amplitude precessional cyclic variations are dominant in the siliciclastic fraction. During the black shale interval enhanced amplitudes of the cyclicity suggests stronger climate contrasts, pointing to alternating conditions with either increased aridity or enhanced humidity. After the black shale event the system remains disturbed for several 10000 yrs before pre-OAE conditions are re-established. The effects of this major perturbation of the climate had also consequences for the biota at the two investigated locations. At the Mazagan Plateau the organic matter production increased even though the upwelling, which had prevailed for more than 200 kyrs prior to the event, weakened probably due to an increased influx of nutrients from the adjacent African Continent. The amount of the bioproduction of the Vocontian Basin remained stable but shifted towards bacterial dominated communities as is indicated by selected isoprenoids (e.g., lycopane). Here, accumulation of organic matter is fostered by enhanced preservation conditions. The entire data record demonstrates a warming pulse during OAE 1b resulting in a weakening of the tropical trade wind system and higher climate contrasts for subtropical regions.

Discussing the trigger mechanisms leading to the abrupt onset and termination of OAE 1b is another important feature of this work. The magnitude of the isotope excursion at the Mazagan Plateau remains constant over several 10000 yrs. By contrast, a single methane blast would strongly affect the atmospheric signal, but the concentration of light carbon dioxide would than constantly exhaust. Hence, the Mazagan record rather suggests a constant supply of ^{13}C-depleted CO_2 during the entire black shale interval. Therefore, a combination of moderate-scale methane pulses supplemented by continuous methane emission at elevated levels over a longer time interval is indicated. The two pulses encompass the total trigger period of OAE 1b, estimated to be about 25000 yrs. The results calculated with a coupled atmosphere–land–ocean–sediment model presented for the Mazagan Plateau reveal similarities to the Vocontian Basin record. Nevertheless, both climate systems reacted differently due to the perturbation at the onset of OAE 1b. These differences can be explained by a shifting Inter Tropical Convergence Zone (ITCZ). Modeled net flux rates, using an Atmospheric General Circulation Model (AGCM), highlight precipitation and evaporation rates for the studied sections during the Cretaceous and highlight strong regional differences dependent on palaeolatitudes. A moderate ITCZ shift of ~15°N during OAE 1b explains the beginning humid climate conditions coupled to alleviative tradewinds at the Mazagan Plateau and also the enhanced climate contrasts at the Vocontian Basin with a general trend towards higher aridity. In summary, the OAE 1b is recorded at several distinctive settings reflecting specific climatic and environmental conditions. These are still identifiable due to characteristic C-isotope curves and temperature proxies. Furthermore, the isochronous widespread deposition and a comparison to literature data indicates that OAE 1b was a global rather than a supraregional event.

Contents

Zusammenfassung ... I
Abstract ... IV

1. Introduction   **1**
   1.1. General conditions during the Cretaceous ... 1
   1.2. Carbon-isotopes, sea surface temperature (SST) and ocean circulation patterns ... 9
   1.3. Oceanic anoxic events (OAEs) ... 13
   1.4. The Lower Albian OAE 1b ... 18
   1.5. Aims of this study ... 22

2. Material and Methods   **24**
   2.1. Samples and preparation ... 24
      2.2. Geochemical analyses ... 25
      2.2.1. Inorganic geochemical measurements ... 25
      2.2.2. Organic geochemical measurements ... 26

3. Study Area   **29**
   3.1. General remarks ... 30
      3.1.1. DSDP Site 545, Mazagan Plateau ... 30
      3.1.2. Vocontian Basin, SE-France ... 34

4. DSDP Site 545, Mazagan Plateau   **37**
   4.1. Palaeogeography and Geology ... 37
   4.2. Results ... 40
      4.2.1. Stable carbon isotopes ... 41
      4.2.2. Organic composition ... 42
      4.2.3. Biomarker ... 46
      4.2.4. Inorganic composition ... 49
      4.2.5. Time model and sedimentation rates ... 55
   4.3. Discussion ... 58
      4.3.1. Conditions prior to OAE 1b ... 58
      4.3.2. OAE 1b at Site 545 ... 64
      4.3.3. The post OAE stage ... 67

5. Vocontian Basin   70

    5.1. Palaeogeography and Geology 70
    5.2. Results 74
        5.2.1. Stable carbon isotopes 75
        5.2.2. Organic composition 77
        5.2.3. Biomarker 83
        5.2.4. Inorganic composition 88
        5.2.5. Time model and sedimentation rates 96
    5.3. Discussion 99
        5.3.1. Conditions prior to OAE 1b 100
        5.3.2. OAE 1b at the Vocontian Basin 111
        5.3.3. The post OAE stage 118

6. Synthesis   121

    6.1. The Trigger of OAE 1b 121
    6.2. Comparison of both sites 127
        6.2.1. Differences and similarities 127
        6.2.2. Regional distinctions and global climate trends 130
    6.3. OAE 1b: Regional or a global event 135

7. References   138

1. Introduction

1.1. General conditions during the Cretaceous

The Cretaceous is in general characterised by extreme climate conditions which were likely unique in the worlds history. Enhanced plate tectonic activity resulted in the breaking up of the super continent Pangaea at the end of the Jurassic. During the Lower Cretaceous the Gondwana breakup led to increased rates of oceanic crust production. High amounts of the greenhouse gas carbon dioxide (CO_2) were consequently released into the atmosphere causing extreme high temperatures, in connection with other conform factors, which than created the so-called Cretaceous "greenhouse world" (e.g. Jenkyns 2003; Huber et al. 2002; Kuypers et al. 1999). In general, a "greenhouse world" is described by elevated volcanism and CO_2 output, increased poleward latent heat transport leading to global warmth, lack of continental ice sheets, highstands of global sea levels, widespread flooding of continental interiors and large-scale global deposition of organic carbon-rich sediments due to oxygen deficiency (Sageman & Holland 1999). The level of atmospheric CO_2 is regulated by the global carbon cycle. To make predictions about past and future climates, the long-term or geochemical carbon cycle is observed which includes interactions between the ocean-atmosphere-biosphere system and the geosphere. Among these major carbon reservoirs of the Earth, the geosphere contains around 1200 times the combined carbon in all surface reservoirs, made up mainly by the burial as carbonate and organic-rich rocks, silicate weathering, CO_2 output from ocean ridges and volcanoes, metamorphism and diagenesis. Fluxes and interactions between the different reservoirs can be simplified in a diagram (Fig. 1) but how feedbacks work and how processes interact can only be estimated.

On a time scale of millions of years the CO_2 level of the ocean-atmosphere system is controlled by the balance between the rate of release of carbon by volcanic activity and the precipitation of carbonates for which cations are supplied by weathering of silicate minerals (Gérard & Dols 1990). During the last 650 kyr atmospheric CO_2 concentrations varied in glacial-interglacial cycles between 180 and 300 ppm and increased nowadays up to 380 ppm in 2005 which is extremely high (Meehl et al. 2007). Atmospheric CO_2 concentrations during the Albian ranged between 500-3000 ppm, depending on the proxy parameter considered (Wagner et al. 2007 and references therein). This record may seem far from recent values but climate studies for the future predict similar values for the year 2100 (Meehl et al. 2007). During the middle to late Cretaceous (110-70 Ma), high rates of organic carbon production and burial, and intense and widespread volcanism had profound effects on the global CO_2 cycle (Arthur et al. 1985). During this time period with high levels of atmospheric CO_2 and associated thermal maxima, the (global) burial of excess carbon across a range of marine environments led to the deposition of the so called oceanic anoxic events (OAEs) (Jenkyns 2003).

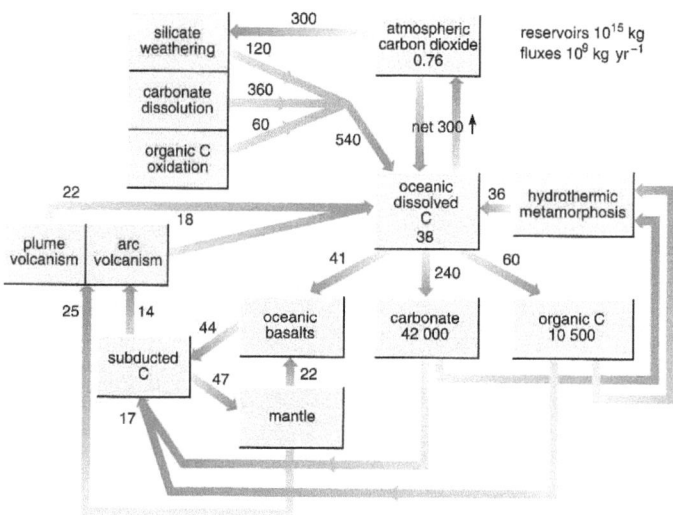

Fig. 1: Box model for the longterm carbon cycle, showing the reservoirs (grey boxes, values given as x10^{15} kg) or supplied by processes (yellow boxes), and estimated fluxes (green arrows, values as x10^9 kg yr^{-1}) of carbon between the reservoirs. Note the large carbon fluxes into the oceans, between the ocean and atmosphere and between the ocean and carbonate reservoirs, and the overwhelming size of the carbonate and organic carbon geological reservoirs (Skelton 2003; after Arthur 2000).

The carbon-isotope record (of the stable isotope ^{13}C) documents several prominent positive and negative excursions during the Cretaceous. Most of them precede or are directly correlated to OAEs (e.g. Jenkyns 2003). The deposition of these organic-rich black shales documents a dominant sedimentary record during the Cretaceous which is unusual because black shale deposition was not limited to local regimes with special oceanic conditions like for example upwelling (Hofmann et al. 1999). They are typical for the

Cretaceous. For further informations see Chapter 1.3.: Oceanic anoxic events (OAEs). Approximately 60% of the known global oil reserves are from the Cretaceous (Irving et al. 1974), in consequence of the higher accumulation rates of organic carbon in deep-sea sediments, with numerous locations in the Persian Gulf and Middle America. An increase of oceanic heat transport and enhanced output of greenhouse gases into the atmosphere are plausible mechanisms to intensify global mean temperatures. But the Cretaceous is not only a time of uniformly warm climates. Although glaciations and polar ice sheets are still highly debated, there were definitely times when global climates were cool or at least cooler than at modern times (Frakes 1999; Steuber et al. 2005). In general, there are times of long-term warming- (Albian-Turonian/Coniacian - earliest Campanian) and cooling trends (Campanian - Maastrichtian) (see Fig. 2; e.g. Huber et al. 2002; Clarke & Jenkyns 1999; Frakes 1999).

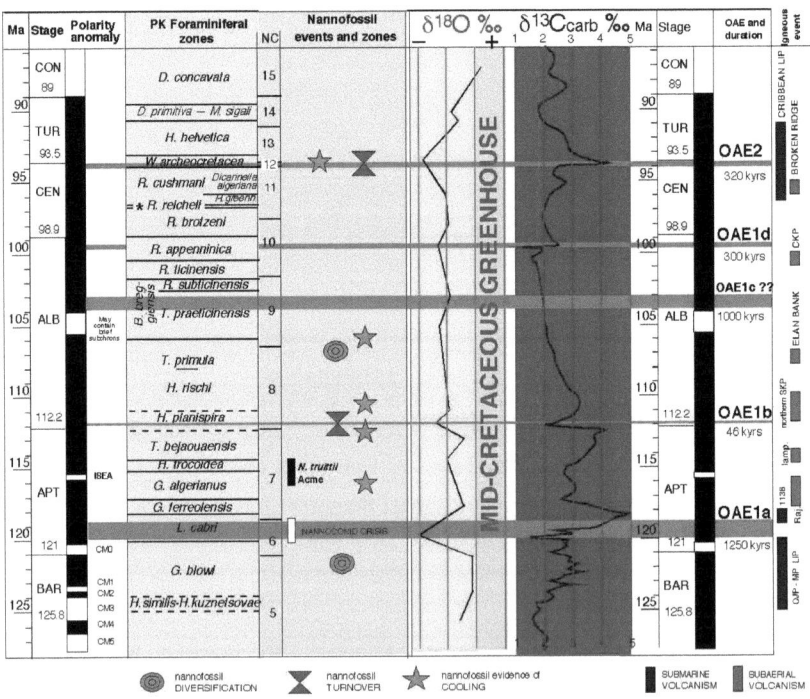

Fig. 2: Mid-Cretaceous integrated stratigraphy, nannofossil evolutionary events and evidence of cooling episodes interrupting the greenhouse conditions. The main igneous events related to the formation of the Ontong Java - Manihiki and Kerguelen LIPs closely correlate with OAEs, isotopic excursions and biotic changes (Bice et al. 2002).

Compared with nowadays, sea-levels were higher, at least temporarily the poles were ice-free and the concentration of carbon dioxide must have been between 3 to 6 or even up to 12 times higher (e.g. Steuber et al. 2005; Bigg 2003; Bice & Norris 2002; Skelton 2003; Kuypers et al. 1999; Barron 1992; Jenkyns et al. 2004). High CO_2 levels are related to the elevated production of oceanic crust in the Lower Cretaceous (Arthur et al. 1985; Larson 1991a&b) as well as to the emplacement of Large Igneous Provinces (LIPs, Larson & Erba 1999) and the catastrophic release of methane by gas hydrate dissociation (e.g. Jahren 2002; Jenkyns 2003; Hesselbo et al. 2000). Climate conditions during the Cretaceous were definitely diversified with extreme hot conditions during the middle part of the Cretaceous (Fig. 3).

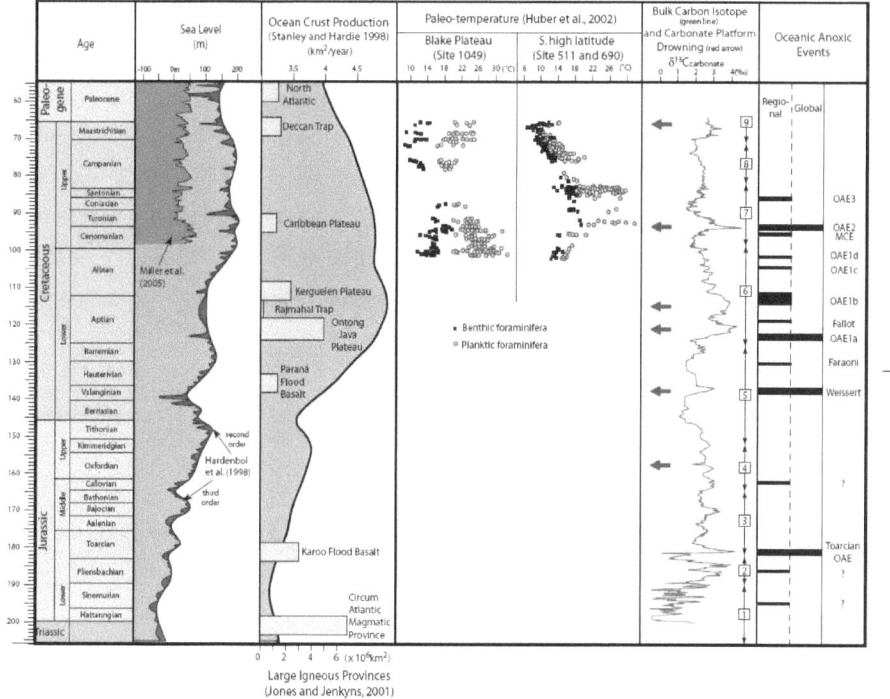

Fig. 3: General conditions during the Jurassic–Cretaceous time period, including changes in sea level, oceanic-crust production, Large Igneous Provinces, palaeo-temperature, bulk carbon isotopes, carbonate platform drowning events and OAEs (Takashima et al. 2006).

INTRODUCTION

A viable alternative to oceanic heat transport during the Albian greenhouse warming is given by increased heat transfer of atmospheric H_2O vapor. Ufnar et al. (2004) showed that the intensified hydrological cycle played a greater role in warming high latitudes than at present times and mass-balance modelling results suggest, that Albian precipitation rates exceeded modern rates at both mid and high latitudes. Ufnar et al. (2004 present zonal profiles of latent heat flux (LHF) values, calculated from precipitation minus evaporation values from modern and Albian mass-balance model simulations.

Evidence for the postulated global warming (and warm polar temperatures) are given by flora, fauna and the record of stable isotopes. Diverse floral and faunal assemblages as indicators for warm temperatures (e.g. coral reefs and breadfruit trees), expanded to high latitudes. Dinosaur areas, assuming that they preferred warm weather conditions, ranged north of the arctic cycle (Crowley & North 1991).

The global warming is verified by analyses of the oxygen stable isotope (^{18}O) record from planktonic foraminifera, indicating that intermediate to deep waters, as well as globally averaged surface temperatures, were about 20°C warm er during the middle part of the Cretaceous than at present (e.g. Bice et al. 2003; Crowley & North 1991 and references therein). Arctic palaeo- sea surface temperatures (SST), measured with the palaeothermometer proxy TEX_{86} (tetraether index of tetraethers with 86 carbon atoms, for further informations see chapter 2.2.2.), also were warmer than 20°C during the mid-Cretaceous (~90 Ma) excluding ice sheets at high latitudes (Jenkyns et al. 2004; Schouten et al. 2002, 2003b). Contrasting evidences indicate polar ice at cooler episodes during the early Cretaceous (Steuber et al. 2005).

Significant evolutionary effects and extinction events affecting the global biota were also present during the Cretaceous. The most dominant is the K/T (Cretaceous/Tertiary) boundary extinction event at the close of the period, marking the end of the dinosaur era. It is the last one of the largest big five Phanerozoic so-called mass extinction events (e.g. Twitchett 2006; Kerr 2001; Hallam & Wignall 1999 and references therein). Although there is obviously no common cause for these events, it is undisputed that all are associated with climatic and/or environmental changes. The K/T extinction event is associated with global cooling as a possible environmental consequence of a cometary impact event, although recent work showed that the impact may have pre-dated the K/T boundary by some 300 kyr (thousand years; e.g. Twitchett 2006). Nevertheless, due to the extreme climatic and environmental situation also major biotic changes in planktonic communities during the Cretaceous time period exist (e.g. Erba & Tremolada 2004).

These biotic changes correspond in general to periods of high accumulation rates of marine organic carbon due to oxygen deficiency (see below). During these phases of higher primary productivity the carbonate content is very low, organic matter is enhanced and the biogenic silica production is increased. A global change in nannofossil-carbonate abundance subjected to these phases, suggest that planktonic communities reacted to changes in trophic and climatic conditions and then persisted during the perturbation phase. Accelerated evolutionary trends are associated with higher fertility and the global warming is triggered by submarine volcanism (Erba & Tremolada 2004; Larson & Erba 1999). Premoli Silva & Sliter (1999) postulated a threefold pattern for the Cretaceous documented by the record of planktonic foraminifers. While the first and second intervals (Valanginian to Aptian/Albian boundary and to latest Albian) are characterised by increasing diversity and complexity, the third interval (end of the Cretaceous) shows rapid diversification and turnover, separated by longer periods of stasis. These extinction/repopulation events are thought to be connected to changes in the physical and chemical structure of the Cretaceous ocean.

Although the Cretaceous stratigraphy is still divided into the Lower and the Upper part, a mid-Cretaceous is unofficially defined by chemical evidences like stable carbon isotopes and carbon-enriched episodes (Ogg et al. 2004). The definition of a time-based mid-Cretaceous period can consequently not be made uniformly. In this work the mid-Cretaceous is defined lasting from 125 Ma (million years ago) to 89.3 Ma, including the stages from Aptian to Turonian after Ogg et al. 2004 (Fig. 4). During this time period the hottest global temperatures for the Cretaceous were reached and anomalously high amounts of volcanic activity were present (e.g. Larson & Erba 1999). The mid-Cretaceous is of special interest because it comprises a lot of detailed information about fauna, flora and rocks as a result of a good geological record. Therefore, conclusions about the environmental setting are feasible and with this information, various theories of climate change are tested (Crowley & North 1991).

The thermal maxima of the Cretaceous "greenhouse world" occurred during the latest Cenomanian and early Turonian, and coolest temperatures were reached during the Maastrichtian, documented by oxygen isotopes from deep-sea sites (e.g. Clarke &

Jenkyns 1999; Wilson et al. 2002; Bice et al. 2003; Huber et al. 2002). In general, the middle part of the Cretaceous is a time period when especially polar temperatures were unusually warm, repeated reef drowning occurred in the tropics and a series of marine carbon-rich sediments were deposited (e.g. Wilson & Norris 2001 and references therein).

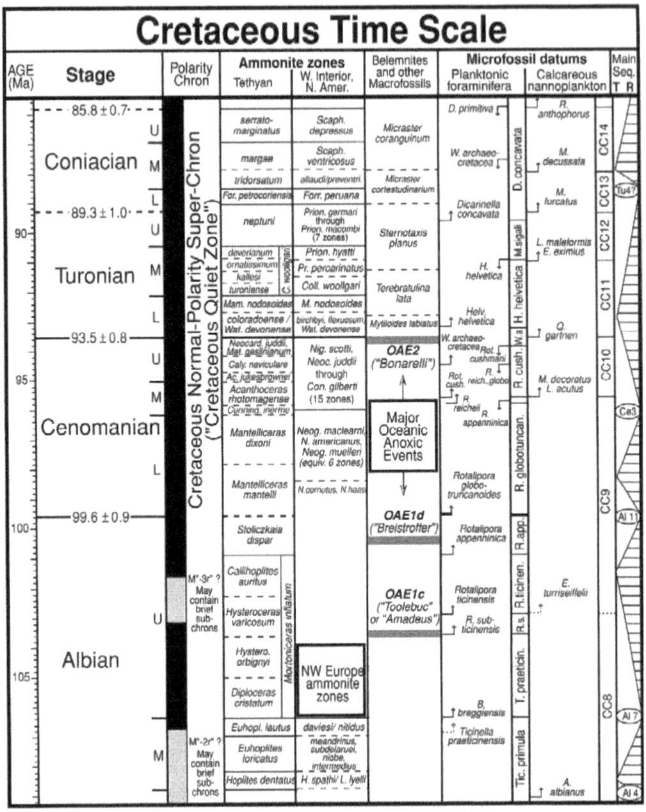

Fig. 4: continued on the next page.

During the Cretaceous shallow continental margins and land surfaces were flooded up to 20% (Bigg 2003; Fig. 5) and the break-up of Pangaea since Late Jurassic times was a cause for the formation of new Ocean basins. Pre-breaking-up of Pangaea, the Palaeotethys Ocean in the eastern part of the supercontinent and the Panthalassa Ocean, the precursor to the Pacific, which covered the rest of the globe, where the two main Oceans. The blocking of possible seaways in the mid-latitudes of the southern and the northern hemisphere resulted in increased western boundary currents transporting heat polewards in the ancient Pacific. The time period comprising the Barremian to Albian (120-110 Ma) represents the onset of seafloor spreading in the North and South Atlantic, and the opening of the Biscay Bay (Schettino & Scotese 2002). Spreading rates were high in the South Atlantic (~43-63 mm/yr) and in the Central Atlantic (55 mm/yr).

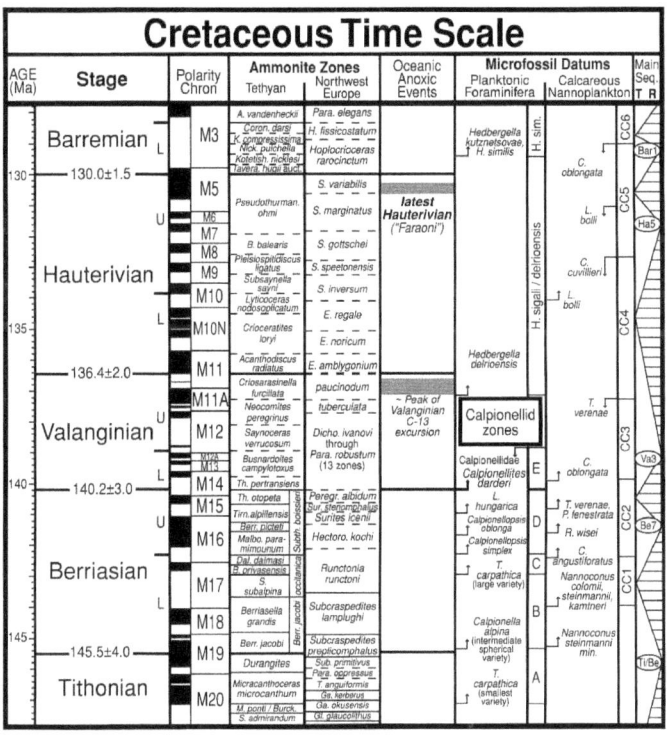

Fig. 4: Stratigraphic time scale for the mid-Cretaceous period (Aptian-Turonian; Ogg et al. 2004).

The temporary globe-spanning tropical Tethys Sea developed as successor of the Palaeotethys (Bigg 2003). By the end of the Early Cretaceous the Pacific, Tehtyan, Atlantic, and Indian Ocean basins separated major continental blocks (Hay et al. 1999). As shown, there are many circumstances that distinguishes the Cretaceous from other time periods. "Greenhouse" global climate conditions due to enhanced submarine volcanogenic events and therewith associated elevated atmospheric CO_2, created an extreme climate scenario. The oceanic heat transport, land-ocean distribution and physical contributions (like expansion and ice melt) were totally different from nowadays. Based on these information, like SST, carbon- and oxygen isotope data, sea-level stands etc., reconstructing the past climate is helpful to make reasonable predictions for the future. For the Cretaceous most cases represent exceptional circumstances compared to modern proportions.

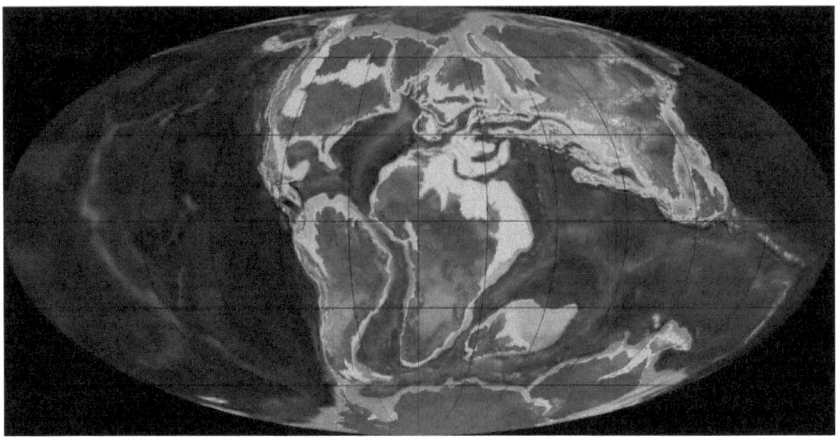

Fig. 5: Mollewide (oval globe) Plate Tectonic Map projection showing global palaeogeography for the late Early Cretaceous (105 Ma). Map created by Ronald Blakey (contact: http://jan.ucc.nau.edu).

1.2. Carbon-isotopes, sea surface temperature (SST) and ocean circulation patterns

The mid-Cretaceous is defined by chemical evidences like stable carbon isotopes and carbon-enriched episodes (Ogg et al. 2004). The two stable isotopes of carbon are called the light and heavy isotopes (^{12}C and ^{13}C) and account for ~98.89 and 1.1 wt.% of all carbon, respectively (Kenneth et al. 2005). The relative abundance ratio of ^{13}C and ^{12}C in a sample is presented as the delta `$\delta^{13}C$´value in parts per mil (‰) defined in equation 1 (after Tyson R.V. 1995). Carbon isotope ratios are principally useful to distinguish between terrestrial and marine organic matter sources in sediments and to identify organic matter from different types of land plants. Organic material is in general depleted in ^{13}C while inorganic material like carbonates are relatively enriched. For organic matter the isotope ratio varies depending on many factors like composition, source and metabolic pathways.

$$\delta\ (‰) = [(R_{sample} - R_{standard})/R_{standard}] \times 1000 \qquad \text{Equation 1}$$

There are various standards used, but still the international standard is the 'PDB' based on a calcite belemnite *Belemnitella americana*, an extinct marine cephalopod from the Maastrichtian Pee Dee Formation of South Carolina, USA. The equation expresses the depletion of ^{13}C relative to ^{12}C in the sample. Modern marine carbonates have an approximate $\delta^{13}C$ value of 0‰, just like the PDB standard. A $\delta^{13}C$ value of −1‰ indicates that the sample has 1‰ less of the heavier ^{13}C isotope relative to the standard and is therefore 'lighter' or better ^{13}C-depleted. Samples with positive values are correlated to the standard enriched in ^{13}C and are termed isotopically 'heavier'. Standard deviations should be in the order of +-0.1-0.3‰. The $\Delta\delta^{13}C$ value is the observed fractionation between the organic matter and the original carbon source. Bulk isotopic data is used to determine the predominant source of suspended or sedimented organic matter. There are mean $\delta^{13}C$ values which are assigned to marine, terrestrial or riverine organic matter.

The carbon-isotope record of the stable isotope ^{13}C documents several prominent positive and negative excursions during the Cretaceous, most of them correlated to global excess carbon burial across a range of marine environments (e.g. Jenkyns 2003). Positive $\delta^{13}C$ excursions are documented across several major Mesozoic carbon-cycle perturbations (Toarcian, Aptian and the Cenomanian/Turonian boundary OAEs) thought to reflect the response of the inorganic carbon reservoir to global burial of ^{13}C- depleted organic carbon (Wilson & Norris 2001). Some of these positive events are preceded by short-term negative $\delta^{13}C$ excursions like the Toarcian (Jurassic), the Aptian and the late Albian/Cenomanian black shale (Wilson & Norris 2001). Rapid changes in the carbon cycle are supposed to be accompanied by turnover in marine biota (see chapter 1.1.).

The carbon isotope record for the Mesozoic is marked by three large positive excursions, identified in all oceans (Larson & Erba 1999). Some are supposed to be linked to superplume events others to OAEs. The carbon isotope curve is considered as a proxy to reflect rates of organic burial. The early Aptian positive excursion, described as a high-productivity global event, is preceded by a short-lived, distinct negative peak which postdates the Selli black shale event (OAE 1a). This negative excursion is not a diagenetic effect because it is documented in various different oceanic sites (Larson & Erba 1999). These records of negative values preceding a positive carbon excursion are interpreted as the result of amplified volcanic activity leading to enhanced CO_2 concentrations in the atmosphere. In the following time period a few more anoxic events are observed during which a negative carbon isotope excursion was documented, one of these is the OAE 1b on which this work is focused.

Sea surface temperatures (SST) as well as temperatures for the intermediate to deep water formations were in general higher, most notably in high latitudes. While SSTs where higher than nowadays at mid- and high latitudes, they were, interestingly, not verifiable for the tropics (Wilson & Norris 2001). In combination with high values of carbon dioxide during the Cretaceous this fact is problematic, because it should also trigger a low-latitude warming (tropical SSTs ~2-5°C warmer than today; Wilson & Norris 2001). Deep waters were warm and saline, generated from a strongly evaporitic environment in the sub-tropics (Bigg 2003). For the Late Cretaceous a good SST record shows significant changes in marine environments while the pre-Albian Cretaceous record is unfortunately conjectural due to a less dense data record (Frakes 1999). There is no good correlation with modern SST gradients and Cretaceous gradients over latitude (Fig. 6).

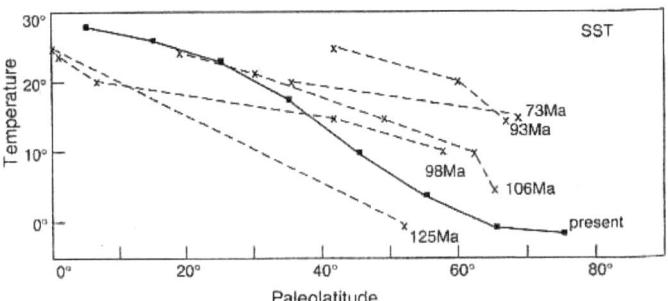

Fig. 6: Modern and selected Cretaceous sea surface temperature gradients (SST) over latitude (Frakes 1999).

In general, the gradients are lower and temperatures are warmer beyond the tropics, with the SST gradient for the Early Cretaceous (125 Ma) as an exception. This SST gradient is most similar to modern conditions, aside from the lack of an extensive record, showing cooler temperatures than nowadays. It contrasts with the flatter and warmer gradients described in all other examples and demonstrates once more that the Cretaceous was indeed warmer than other time periods, but was also characterised by cold/warm fluctuations as explained before (see Chapter 1.1.). A problem is that all temperature gradients could be influenced by undetected salinity variations. Higher salinity, effected by evaporation and continental runoff, would lead to more positive $\delta^{18}O$ values, interpreted wrongly as lower temperatures. Significant palaeoclimate- and palaeoceanographical conditions during the Cretaceous comprise atmospheric heat transport, the wind system, warm and cold water masses and salinity. During the Albian, a zone of high salinity (>37ppt) established over the North Atlantic between 15°N and 35°N latitude (Frakes 1999 and references therein; Barron 1992). Salinity changes have relativeley more impact on the density of cold water and warmer water is affected by changes in temperature (Bigg 2003). Usually the winds weaken during warmer periods due to decreased thermal gradients. Nevertheless, calculated Cretaceous simulations show that the wind systems were increased which is not surprising looking at the latent heat released by tropical convection. Higher tropical SSTs would lead to higher levels of latent heat added to the atmosphere. The oceanic conveyor belt, like it is known nowadays, transports huge quantities of stored heat from south to north in the Western Atlantic and Pacific Ocean (Fig. 7). It is basically driven by surface wind patterns and differences in water density (salinity) and temperature, leading to upwelling and mixing of warm and cold water masses. These density differences drive the large-scale ocean currents formally known as a thermohaline circulation.

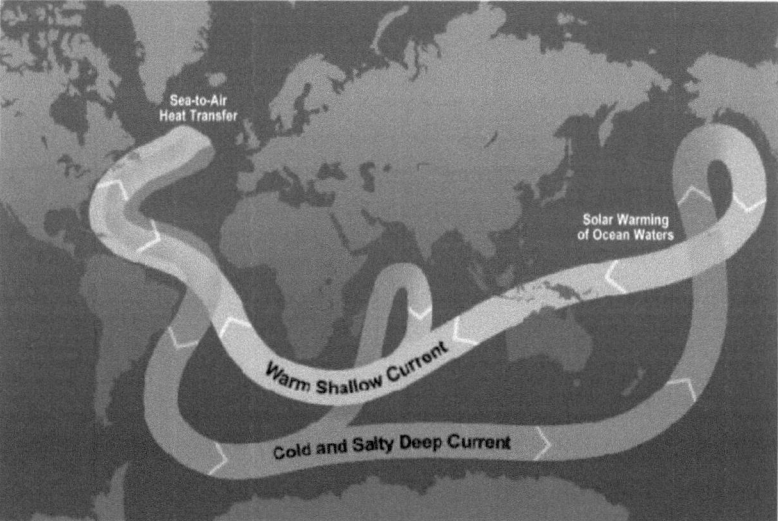

Fig. 7: The present oceanic conveyor belt in a simplified illustration. It is driven by differences in heat and salinity (http://www.usgcrp.gov/).

Changes in salinity concentrations would lead to a different heat transport. In the Cretaceous, due to the opening ocean gateways, some ocean basins were shallower than today, leading to missing cold and/or more saline, deep currents which are relevant for receiving a closed "mixing" system. A strong thermocline, comparable to the modern one, was first developed in the late Albian. Initial stratification of the Cretaceous water masses is explained by an increase in the Earth's latitudinal thermal gradient (Premoli Silva & Sliter 1999).

It is accepted that significant export of heat from the tropics took place by ocean currents (Frakes 1999). During the Cenomanian, there was probably no deep-water exchange between the proto-North Atlantic and the Southern Ocean. The North Atlantic was a rather isolated ocean basin. Simulations for the Turonian document a time of increasing surface water exchange across shallow seas and possibly an opening deep gateway between the North and South Atlantic (Pletsch et al. 2001). According to Voigt et al. (2004 and references therein) the formation of oceanic deep waters by cooling warm and saline water took place in high latitudes of the northern pacific and southern hemisphere. During the Early- and mid Cretaceous the North Atlantic was rather a restricted ocean basin. The exchange of global deep water masses was very limited. Slow ocean circulation and limited ventilation of intermediate and deep waters are assumed for mid-Cretaceous oceans (Wilson & Norris 2001).

The thermal maximum of the Mesozoic-Cenozoic greenhouse climate was reached during the late Turonian, with tropic SSTs around 33-34°C and southern subpolar Atlantic Ocean temperatures around 30-32°C (Voigt et al. 2004).

1.3. Oceanic anoxic events (OAEs)

Oceanic anoxic events were first defined by Schlanger & Jenkyns (1976). Accordingly, black shales are marine products of anoxic events as their deposition was not restricted by local basin geometry. These organic-rich deposits are typical indicators of warm climate conditions. Their formation was restricted locally and temporally to specific periods and environments of Earth's history. Formation was focused on Cretaceous times with the except of the early Toarcian OAE (e.g. Jenkyns 2003; Erba 2004). Black shales are dark coloured, mostly laminated, organic carbon-rich (oc-rich) and bituminous marine sediments confined to short time intervals. Their chemical composition, primarily the types of organic matter, as well as their thickness, varies drastically. Their occurrence was explained by the development of a widespread and thick O_2-minimum zone (OMZ) and poor oceanic mixing in the world oceans (Schlanger & Jenkyns 1976). The fact, that they are globally distributed and deposition took place in a variety of palaeo-bathymetric settings, not restricted to local basin geometry, led to the definition of "Oceanic Anoxic Events" (Schlanger & Jenkyns 1976; Jenkyns 1980).

Two features favoured their formation: the higher sea-level stand leading to an increase of organic carbon production, and the warmer global climate leading to diminished oxygenated bottom water formation (Schlanger & Jenkyns 1976). Schlanger & Jenkyns (1976) describe the deposition of OAEs as a consequence of the interaction of transgression and the plate-tectonic framework and all hence resulting factors during the Cretaceous. In addition to the palaeogeographic and –oceanographic development during Cretaceous times (see Chapter 1.1.) this time period is accompanied by rapid sea floor spreading and an increase in the volume of the mid-ocean ridge system, which led to a major transgression of the sea, flooding the low-lying coastal plains. Furthermore, the burial of organic carbon, SST and the sea level were much higher compared to modern or any other time in Earth´s history (e.g. Arthur et al., 1990; Larson 1991a, b; Fig. 8).

Three dominant OAEs are documented for the Mesozoic: the Toarcian, the Early Aptian (OAE 1a) and the Cenomanian-Turonian Event (OAE 2; e.g. Erba 2004; Jenkyns 2003; Jenkyns 1980). These events are seen as global, as the OC-rich layers can be correlated over large distances (Schlanger & Jenkyns 1976) and their appearance is world wide. The second event was further divided into the OAE- 1b (early Albian), -1c (late Albian) and -1d (late Albian-Cenomanian). They display more regional depositions and cover a shorter time period. Nevertheless, their distribution can also be wide spread and the term of regional can in some cases be extended to a supraregional black shale event (Friedrich et al. 2005). The final Cretaceous event is the Coniacian/Santonian OAE 3 which was restricted to continental marginal settings and is not of global nature (e.g. Beckmann 2005; Leckie et al. 2002).

INTRODUCTION

Following the first definition of OAEs by Schlanger & Jenkyns (1976), several models were established to explain their origin and spatial distribution in more detail. One classical model proposes the sedimentation of black shales in addition to an expansion of the OMZ in restricted, stagnant ocean basins leading to increased rates of organic carbon burial (e.g. Schlanger & Jenkyns 1976; Bralower & Thierstein 1984). Another model is based on sedimentation primarily driven by high organic productivity, induced by upwelling (e.g. Erbacher et al. 2001). In addition, Erbacher et al. (1996) distinguished sea-level controlled models: The P-OAE (productivity oceanic anoxic event) as the result of increased oceanic productivity and the D-OAE (detrital oceanic anoxic event) as the result of increased sedimentation rates of terrestrial organic matter. Furthermore the connection between the release of methane from gas hydrates, contained in marine continental-margin sediments, and the deposition of raised marine organic carbon became more evident, at least for some black shale events (Fig. 9; e.g. Jenkyns 2003).

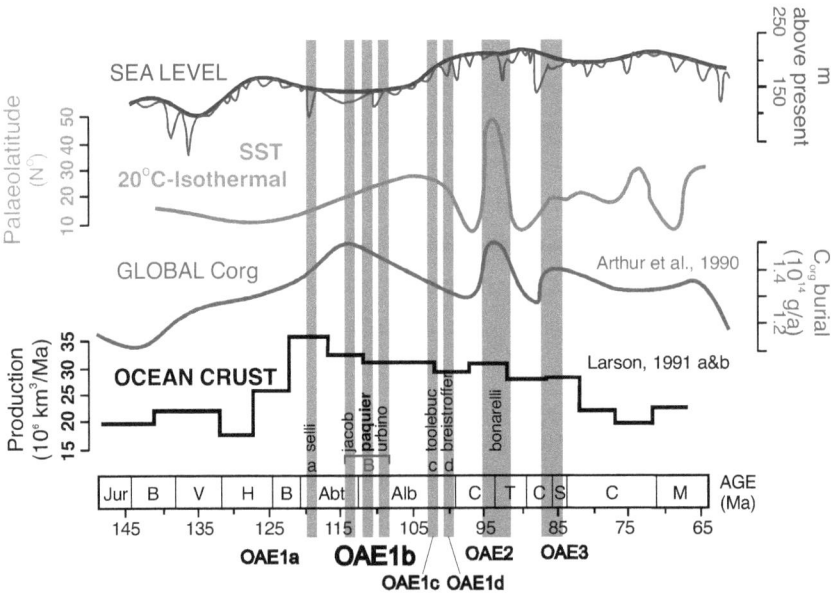

Fig. 8: General conditions during the Cretaceous including the OAE intervals (after Beckmann 2005).

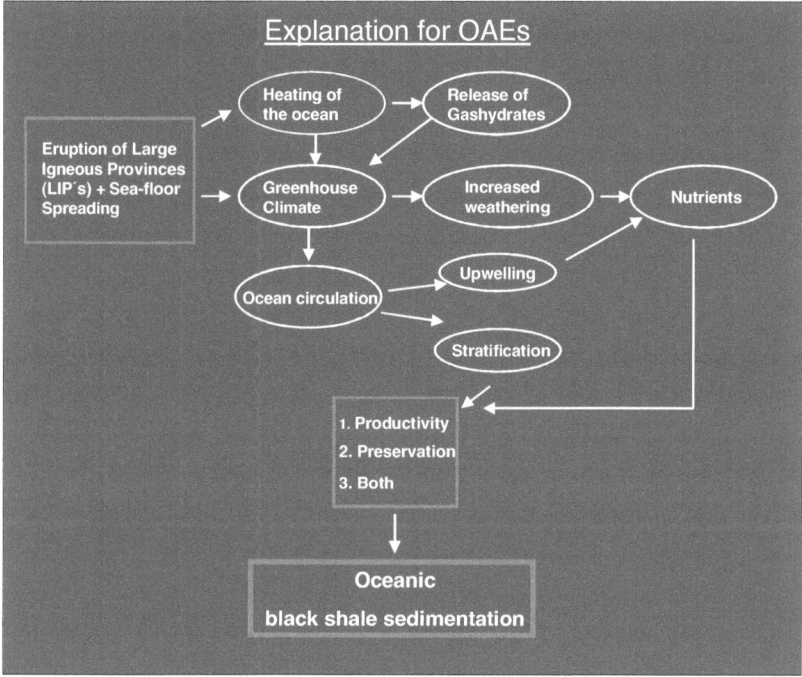

Fig. 9: Simplified diagram displaying the linkage of OAE sedimentation to tectonic and/or climatic disruptions and the resulting conditions. The potential role of methane gas hydrates in the global carbon cycle is illustrated. Dissociation of gas hydrates, related either to warming of the bottom waters and/or tectonic disruption of the sedimentary column, leads to release of methane gas.

OAEs are primarily characterised by high accumulation rates of organic carbon and an isochronous deposition. In general, their appearance is also correlated to positive $\delta^{13}C$ excursions (e.g. Erbacher et al. 1996; Kuypers et al. 2002; Leckie et al. 2002) which are caused by the burial of ^{13}C-depleted OC and, therefore, an enrichment of ^{13}C isotopes in the worlds oceans. But this assumption appears to be simplified looking at the very different "general" conditions for each OAE. A negative carbon isotope excursion preceded global excess carbon leading to the formation of the Toarcian OAE and the early Aptian OAE 1a (Jenkyns 2003). This negative shift in carbon- and oxygen isotope ratios is seen as the consequence of dissociation, release and oxidation of gas hydrates from continental-margin sites (Jenkyns 2003). The introduction of light carbon into the ocean-atmosphere system is accompanied by extreme global warming by a few degree in only a few thousand years and followed by an abrupt positive $\delta^{13}C$ excursion due to increased marine productivity, black shale deposition and sea-level rise (Erba 1994; Föllmi et al. 1994). These anoxic events are thought to be caused by high productivity and export of carbon from surface waters.

The OAE 1b is marked by a distinct negative carbon isotope shift which is not followed by a clearly positive $\delta^{13}C$ isotope shift (e.g. Herrle et al. 2004; Erbacher et al. 2001). In contrast to high productivity due to vertical mixing of intermediate and surface waters, Erbacher et al. (2001) proposed, that the formation of OAE 1b was controlled by variations in precipitation and/or continental runoff as a mechanism for changing deep water convection. Decreasing $\delta^{18}O$ values are related to a drop of surface water salinity and minor increases in surface water temperatures due to increased continental runoff at warm periods. Erbacher et al. (2001) define the formation of OAE 1b as a "super-sapropel stage", reflecting the higher geographical extent and the longer duration according to the Quaternary Mediterranean sapropels. The late Albian OAE 1c ("Toolebuc" event − 101 Ma) has a more regional significance without indications of planktic protozoa extinction and increased productivity (Erba 2004, Erbacher et al. 1996). Erbacher et al. (1996) described this event as the result of a lowstand system tracts and named these levels D-OAEs. The distribution of the latest Albian OAE 1d ("Breistoffer" event − 99,5 Ma) is geographical very widespread although mainly found in the opening Atlantic and the Tethys ocean (after Wilson & Norris 2001). In contrast to the "mega-sapropel" OAE 1b, the Breistoffer–Event represents a perturbation to the global carbon cycle with a pronounced positive $\delta^{13}C$ isotope shift during the main phase of black-shale deposition (Wilson & Norris 2001). The authors found similarities to the three global deposition events (Toarcian, Early Aptian and Cenomanian) and describe OAE 1d as a global event. It was triggered by a collapse of upper water column stratification due to surface water cooling or an intermediate water warming event (Leckie et al. 2002).

Nevertheless, the Cenomanian/Turonian Boundary Event OAE 2 ("Bonarelli Event − 93,5 Ma) is the most widespread (Leckie et al. 2002) and truly global in nature (Erba 2004), with marine black shales deposited in all major oceans and epicontinental seas and a pronounced positive carbon-isotope excursion in all organic and inorganic carbon reservoirs (Jenkyns 2003). During OAE 2 major turnover (extinction followed by origination) within calcareous nannofossils occurred (e.g. Erba 2004, Leckie et al. 2002). OAE 2 was not preceded by a negative $\delta^{13}C$ shift, indicating the absence of gas hydrate dissociation and indicating that methane is not necessary for oceanic anoxia. This event was likely caused by an expanded oxygen minimum zone associated with elevated productivity and/or by the decay of the thermocline due to an abrupt deep-sea warming event (Leckie et al. 2002). OAE 1a and OAE 2 are both associated with increased marine productivity and submarine volcanism. The last OAE during the Cretaceous is the OAE 3 (86-85 Ma). The deposition of black shales occurred regional in low latitudes with restricted ventilation like sheltered subbasins along the upper continental margin and is highly influenced by a dynamic Cretaceous climate-ocean system (Wagner et al. 2004).

OAEs were extraordinary events, driven by extraordinary forcing factors. As mentioned, each was associated with pronounced isotopic excursions and elevated rates of planktonic turnover (Leckie et al. 2002). Leckie et al. (2002) conclude that all mid-Cretaceous black shales were linked to submarine volcanism and that the emplacement of oceanic plateaus and increased ridge crest hydrothermal activity helped indirectly sustaining elevated levels of marine productivity by iron fertilization of the water column. Although OAEs have been studied for more than 30 years, their origin and spatial distribution is still under debate.

1.4. The Lower Albian OAE 1b

A supraregional versus global nature of OAE 1b is controversely discussed in the literature. While most authors favour a supraregional Event, some argue for a global Event comparable to the early Aptian OAE 1a (e.g. Erbacher et al. 1998; Bralower et al. 1999; Herrle et al. 2004). OAE 1b is documented in Albian deposits from the North and Central Atlantic region as well as from the western Tethys area. Friedrich et al. (2005) were able to correlate the strongest benthic foraminiferal repopulation event among all their investigated sections which include the Vocontian Basin, the Mazagan Plateau and Blake Nose in the Atlantic Ocean, and infer a supraregional distribution. Lasting approximately 45 ka, based on orbitally tuned time models, OAE 1b has the shortest duration amongst the Cretaceous black shales (Hofmann et al. 2008 and references therein; Erbacher et al. 2001; Herrle et al. 2003) the remainder of which have lasted from 160 ka (OAE 1d) up to 1 Ma (OAE 1a). The Valanginian Weissert Event is thought to have lasted up to 2.3 Ma (Sprovieri et al. 2006). The OAE 1b was furthermore accompanied by elevated carbon burial although the amount of carbon was moderate in comparison to other OAEs in the Cretaceous (Wagner et al. 2007). The processes that led to its deposition and the related consequences for the global atmosphere ocean system and the biotic evolution are presently highly debated. Importantly, major transitions occurred in tectonics, sea level, climate, lithofacies and marine plankton communities (Leckie et al. 2002). Black shale horizons deposited during Aptian to Albian times are in general linked to sea level changes (e.g. Bréhéret 1994a; Erbacher et al. 1996, 1999).

The deposition of black shales comprising the oceanic anoxic event 1b began in the late Aptian and ended in the early Albian (e.g. Bréhéret 1994a, Erbacher et al. 1996, 1998, 1999; Erbacher & Thurow 1997; Herrle 2002; Leckie et al. 2002). It comprises four prominent events; the uppermost Aptian black shales ("Niveau Jacob", "Niveau Kilian" in the Vocontian Basin in SE-France, and the "Livello 113" and "Monte Nerone" in central Italy) and the lower Albian black shales ("Niveau Paquier" and "Niveau Leenhardt" in the Vocontian Basin, and the "Livello Urbino" event in the Apennines of central Italy; Leckie et al. 2002; Herrle 2002; Fig. 10). The uppermost Aptian black shales ("Jacob" and "Kilian") were separated from lower Albian black shales ("Paquier" and "Urbino") by a sea level fall near the Aptian/Albian boundary (Weissert et al. 1998; Bréhéret 1994a). Also the TOC content differs drastically; the "Jacob" as well as "Kilian" and the latest "Leenhardt" do not exceed values above 3.3% TOC, while the "Paquier" event show TOC values up to 8% and is identified as a very finely laminated paper shale (Bréhéret 1994a).

The "Jacob" event represents deposition in a sea-level low stand system which resulted in increased detrital organic matter supply and burial (Bréhéret 1994a). Erbacher et al. (1998) described the "Paquier" as well as the latest "Leenhardt" event as productivity events (P-OAE) consisting mainly of the kerogen type II reflecting a marine origin (Bréhéret 1994a). The "Paquier" and the "Urbino" black shale events are seen as the regional equivalent of OAE 1b (e.g., Erbacher et al. 1998, 1999, 2001; Leckie et al. 2002, Herrle 2002, Herrle et al. 2003). Bréhéret (1985) described equivalent sections in South Germany near Hohenschwangau ("In der Höll") and from Austria in the section "Losenstein". A prominent third site from the Tethyan realm is the uppermost Vigla black shale section of the Ionian Zone in NW-Greece (Tsikos et al. 2004). Bralower et al. (1999) specified this black shale exposed in ranges of the Sierra Madre Oriental fold and thrust belt of the northern Mexican Cordillera. OAE 1b is described from several Ocean Drilling Program (ODP) / Deep Sea Drilling Program (DSDP) legs from the Atlantic.

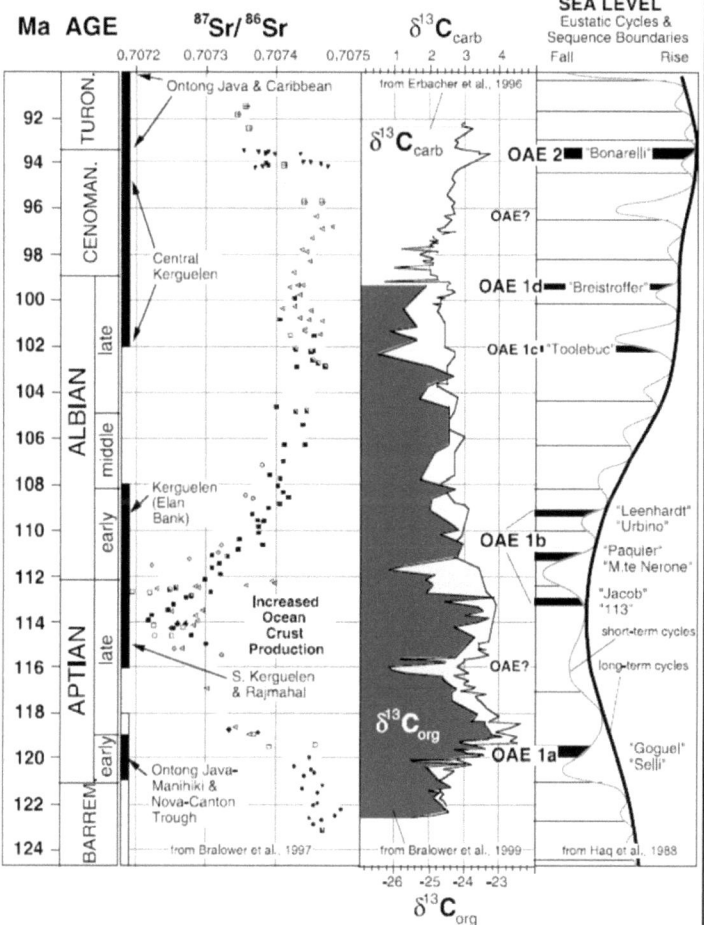

Fig. 10: The mid-Cretaceous record of major black shales and oceanic anoxic events (OAEs) in the context of the carbon isotopic record, changing global sea level and seawater chemistry. Short-term sea level changes are shown as the dark shaded line and the long-term record of sea level is shown with the thick solid line. OAE1a in the early Aptian ("Selli event"), OAE 1b spanning the Aptian/Albian boundary (including the "Jacob," "Paquier," and "Urbino" events), and OAE 2 at the Cenomanian/Turonian boundary ("Bonarelli event") each correspond to negative $^{87}Sr/^{86}Sr$ excursions indicative of elevated submarine volcanism. Initiation of increased spreading rates drove the long-term (Albian-Turonian) rise of global sea level. After rising to a plateau by the late Albian, perhaps due to increased rates of continental weathering with global warming, the $^{87}Sr/^{86}Sr$ ratio once again dropped sharply at or near the time of OAE 2 (Leckie et al. 2002, and references therein).

These are e.g.: DSDP Site 545 at the Mazagan Plateau, ODP Site 641 at the Galicia Bank, DSDP Site 511 at the Falkland Plateau, DSDP Site 386 at the Bermuda Ridge, ODP site 1276 at the Newfoundland Basin and ODP Site 1049 at the Blake Nose Plateau (Fig. 11; e.g., Bralower et al. 1993; Erbacher et al. 1999; Leckie et al. 2002, Herrle 2002, 2003; Shipboard Scientific Party Leg 210, 2004). Based on biogenic data from DSDP Site 545 (Leckie 1984), Leckie et al. (2002) described OAE 1b as a high productivity event. The Albian black shale is dated to the *planispira* planktonic foraminiferal zone and the *tardefurcata* ammonite Zone (Herrle 2002; Herrle et al. 2003) with a likely absolute age of 111 Ma, after the time-scale of Gradstein et al. (1994).

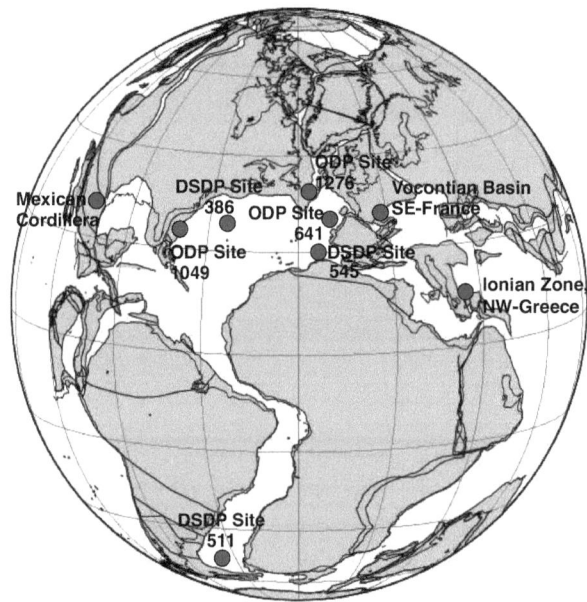

Fig. 11: A map of the Albian (112ma) showing some of the most prominent known OAE 1b deposits of the Atlantic- and Tethyan realm (Palaeomap reconstruction after Hay et al. 1999).

The causes of the perturbation during the deposition of OAE 1b still remain unclear (Hofmann et al. 2008). Orbitally forced mechanisms (precessionaly controlled monsoonal forcing) leading to the depositions in the Tethyan realm (here the Vocontian Basin) were proposed by Herrle et al. (2003). The described climate signal and concurrent productivity changes are regionally restricted to low latitudes. High monsoonal forcing led to increased humidity and wind stress, resulting in high terrigeneous input and high surface water productivity. The climate during OAE 1b is interpreted to reflect extreme warm and humid conditions which is also documented from the Blake Nose and the Mazagan Plateau (Erbacher et al. 2001; Herrle 2002) by increasing surface temperatures. Beside changes in productivity, Herrle et al. (2003) speculated that diminished deep water formation, leading to a decrease in ventilation an thus to enhanced preservation conditions, were the two factors triggering OAE 1b formation.

This assumption is similar to the one proposed by Erbacher et al. (2001). They analysed OAE 1b at ODP Site 1049 with a planktic and benthic foraminiferal stable isotopic data set (see chapter 1.3.) and their results show increasing oxygen isotope gradients from surface to deep-waters and, hence, suggest black shale deposition is forced by a reduction in ventilation of subthermocline waters. A gradual reduction of vertical temperature and salinity gradients, results in a thermohaline stratification of the water column, which is accompanied by surface water warming and/or increased runoff, caused formation of OAE 1b rather than high productivity. Furthermore it is characterised by a distinct negative shift in various carbon reservoirs (Wagner et al. 2007) and by the composition of the organic matter which is up to 80 weight percent derived from marine, nonthermophilic archaea living chemoautotrophically (Kuypers et al. 2001, 2002; Tsikos et al. 2004).

The study presented here focusses on the OAE 1b located at the DSDP Site 545 at the Mazagan Plateau, and the Niveau Paquier/OAE 1b at the Vocontian Basin in SE-France.

1.5. Aims of this study

The deposition of the Lower Albian OAE 1b was clearly a major disturbance of the carbon cycle and ensuing, of the carbon isotope balance, because of the excess burial of organic carbon. The onset of OAE 1b is furthermore accompanied by the already mentioned distinct negative excursion in various carbon isotope reservoirs (Wagner et al. 2007, 2008). Negative isotopic carbon excursions are also documented, e.g. for the Toarcian OAE and the Early Aptian OAE 1a which are thought to be possibly caused by the rapid release of methane from marine gas hydrates, destabilised by warming bottom water induced by inflowing volcanogenic CO_2 (e.g. Jenkyns 2003; Hesselbo et al. 2000; Jahren et al. 2001; Larson & Erba 1999). These scenarios are accompanied by rapid global warming and major perturbations of the global atmosphere-land-ocean system.

Mechanisms leading to the deposition of OAE 1b, as well as the initial trigger are highly debated. The main objectives of this study are to highlight dominant processes and feedbacks before, during and after Albian black shale deposition including continental input, wind-systems, ocean currents as well as the composition and origin of organic matter. Preliminary studies showed that sediments of the OAE 1b are widely spread over the North- and Central Atlantic and the Tethyan realm. The intention was to generate broad and precise geochemical data sets from two differing sites, which allow for general statements on contrasting climate-systems and which show possible differences in local settings. Regional effects are visible in an overall climate system and illustrate differences in global events. Therefore, high-resolution geochemical records for two palaeo-geographically distinct sites were established. Detailed organic and inorganic geochemical analyses were performed on a 6 meter record from DSDP Site 545 (North Atlantic, Mazagan Plateau) and a 14 meter record from the Vocontian Basin (SE-France). DSDP Site 545 was located in the lower Albian northern tropical climate belt offshore Morocco. The Vocontian Basin was situated in the subtropical Albian climate belt in the western Tethys region.

By linking millennial-scale geochemical records from DSDP Site 545 at Mazagan Plateau off subtropical NE-Africa and sections from the Vocontian Basin in the western Tethys to global atmosphere-land-ocean models, the main processes and feedbacks of OAE 1b are shown. DSDP Site 545 and the Vocontian Basin were chosen as sediment records, because these sites were already studied in detail with both biostratigraphic and carbon isotope stratigraphic methods (e.g. Leckie 1984; Herrle 2002). Hence, Site 545 is one of the best characterised section containing OAE 1b. The here presented new approach comprises multi-proxy measurements yielding sea surface temperature (SST), specific biomarkers and the composition of the terrigenous input by specifying the inorganic material.

2. Material and methods

2.1. Samples and preparation

DSDP Leg 79, Site 545 was drilled in 1981 on the Mazagan sector in the Central Atlantic. The cores are stored in the repository at the Lamont-Doherty Earth Observatory, Columbia University in NY, USA. From a 6 meter long record (394,5-388,5 mbsf), including the 80 cm thick black shale, samples were taken in 1 cm-increments, yielding a total around 600 samples, with a maximum time resolution in the order of 700-800 years per sample (Erbacher et al. 1999; Herrle 2002).

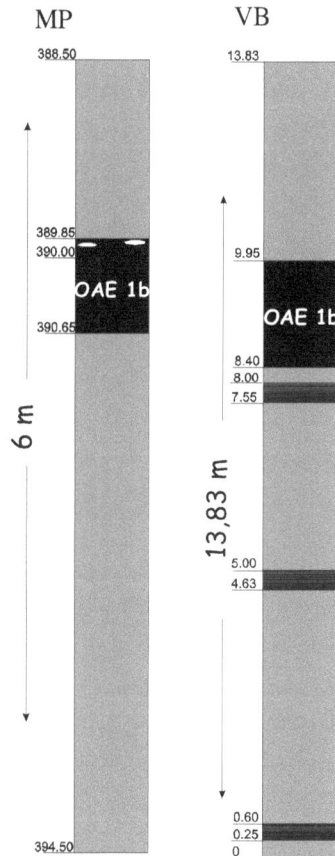

Fig.12: Simplified profiles for both analysed sections. Left, Site 545 Mazagan Plateau (MP) and on the right the Vocontian Basin (VB) section. Note, the VB section is more than twice as long as the MP interval.

The material from the Mazagan Plateau is in general composed of greyish claystones and clayey chalks and the black shale facies (Leckie 1984). The material from the Vocontian Basin was obtained during three field trips (by Jens O. Herrle) in 1998 and 1999 and is stored at the University of Tübingen. The L'Arboudeysee section comprising the Niveau Paquier black shale (sensu stricto OAE 1b) is a 14 meter long record in intervals varying from a few (within the black shale) up to ten centimeter in densely. The record of 14 m of marlstones yielded a total of around 200 samples. The marlstones are punctuated by the black shale layers Haut Noir 7 (HN7), Haut Noir 8 (HN8; Herrle 2002) and by the black shale Niveau Paquier reaching about 1,55 metres in thickness. A third OM-rich layer exist just below the OAE 1b. The focus of this study is the OAE 1b, the other black shale layers are not further discussed.

All whole rock samples were taken to the laboratory at the University of Cologne. After a detailed core description identifying lithology, bioturbation and lamination, all samples were oven-dried at 35°C and homogenized prior to further analyses. Measurements were conducted in Cologne, the University of Bremen, the University of Newcastle upon Tyne (UK), and at the Royal Netherlands Institute for Sea Research (NIOZ) in NL.

2.2. Geochemical analyses

2.2.1. Inorganic geochemical measurements

The aim of this study was to highlight main continent-ocean processes and feedback mechanisms prior, during and after the black shale deposition. Predictions about the composition of the terrestrial input and highlighting wind- and ocean currents were anticipated to reconstruct a Cretaceous climate scenario. Analysing the inorganic composition was a step forward to achieve this aim. Hence, the inorganic sediment chemistry was determined by wavelength-disperse X-ray fluorescence spectometry (XRF) on the homogenized sample and the bulk carbon isotopes were specified in high resolution.

Bulk carbonate isotopes for both sites were analyzed at the Southampton Oceanography Centre (Herrle 2004) using a Europa Geo 20-20 mass spectrometer equipped with a automatic carbonate preparation system (CAPS). Isotope analyses were therefore extended on the existing record, choosing a closer sampling interval. The organic carbon isotopes for $\delta^{13}C$ were determined on decalcified samples in Cologne and at the NIOZ. After combustion at 1050°C, the resulting CO_2 was trapped and analysed with a Finnigan MAT Delta E mass spectrometer. All $\delta^{13}C_{org}$ values are reported relative to the V-PDB standard. Data points presented are within error bars and both corrected for blanks and standards. A record of nearly 70 additional isotope values was newly generated for the Vocontian Basin (VB) as well as for the Mazagan Plateau (MP).

To testify the terrestrial aeolian dust input, the major-element sediment chemistry was determined by wavelength-disperse X-ray fluorescence spectometry (XRF) on the milled sample. Therefore the powdered sample were heated first for four hours at 105°C. The loss on ignition (LOI) is determined on an aliquot by heating to 1000 °C for 120 min. The residue is mixed with lithium-tetraborate and fused for 30 minutes at 1200 °C. Glass beads were produced and analyzed using a sequential X-ray spectrometer (Phillips PW2400) calibrated with natural and synthetic standards. The concentrations of 30 major and trace elements were determined. Predictions about the clay mineral composition and the quartz fraction were feasible and gave therefore information about ancient climate systems, because distinct source areas can be discriminated by variations in the abundance of quartz and illite versus clay-mineral content (smectite, kaolinite, illite; Wagner & Pletsch 1999). Samples were measured at the University of Cologne with an average spacing of 3 cm for the MP and on the total record from the VB yielding a total of around 200 for each site.

2.2.2. Organic geochemical measurements

In order to identify the organic matter, high resolution analyses were conducted to classify source and origin particularly of the black shale. Bulk organic isotopes, total sulphur (S_{tot}) content and TOC contents were determined by decalcifying powdered rock samples with 2N hydrochloric acid, analysing the decalcified sediments in duplicate on a Carlo Erba 1112 Flash Elemental Analyser coupled to a Thermo-Finnigan Delta Plus isotope mass spectrometer (Royal NIOZ) and a LECO CS 300 (Bremen). Analytical errors for TOC are below 3%, and for $\delta^{13}C_{org}$ (‰ versus VPDB) are 0.1 ‰. Standard external analytical precision, based on replicate analysis of in-house standards to NBS-19, is better than 0.1‰ for $\delta^{13}C$.

Pyrolyses of organic matter (hydrogen index, HI; oxygen index, OI) was conducted on a Rock Eval 3-instrument and Rock-Eval 6 with S3, according to the method described by Espitalié et al. (1984) on powdered whole rock samples with a TOC-content of more than 0.3% following the suggestions of Peters (1986). The average sample spacing for Rock Eval (RE) analyses was up to every 2 cm for intervals with high OC content and approximately every 10-15 cm for intervals with low OC content. Around 80 samples were analysed with RE for the Mazagan Plateau and up to 25 for the Vocantian Basin.

Bulk organic isotopes were supplemented by measuring the compound-specific stable carbon isotope data from marine C_{27}-sterenes and terrestrial long-chain C_{27}, C_{29}, C_{31} n-alkanes and TEX_{86}-sea surface temperature (SST in °C; see Wagner et al. 2008). For quantifying the biomarkers, the composition of the organic material was determined by gas chromatography-mass selective detector- (GC-MSD) and isotope ratio monitoring- (IRM) analyses. N-Alkanes, Steranes and Sterenes, Hopanes and Sterolethers were detected as the aliphatic fraction. Short chained n-alkanes mainly indicate aquatic sources, while long chained n-alkanes are derived from higher plant waxes. Steranes and the unsaturated Sterenes derive from the cell membrane of eukaryotic organisms while hopanes derive only from prokaryotic sources. The abundance of terrestrial input and sources of organic matter was calculated to support the discussion using the long- vs. the short-chain n-alkanes ratio displayed in the *carbon preference index* (CPI), a measure of thermal maturity and dominance of terrigenous organic matter using nC_{24} to C_{32} n-alkanes (Bray and Evans 1961). In most samples the n-alkanes nC_{27}, nC_{28}, nC_{29} and nC_{31} are dominant, indicating a significant terrestrial source.

To quantify the TEX_{86} SSTs, freeze-dried and homogenized sediments (~1.5 g dry mass) were extracted with dichloromethane (DCM)/methanol (2:1) by using the Dionex accelerated solvent extraction technique. The extracts were separated by Al_2O_3 column chromatography using hexane/DCM (9:1), DCM/methanol (95:5), and DCM/methanol (1:1) as subsequent eluents to yield the apolar, glycerol dialkyl glycerol tetraether (GDGT) and polar fractions, respectively. The apolar fractions were analysed by gas chromatography (GC) and GC-mass spectrometry. Prior to analyses by GC-isotope ratio monitoring spectrometry, the apolar fraction was separated with an $AgNO_3$ column using hexane and

DCM to yield the aliphatic and the non-aliphatic fractions, respectively. Perdeuterated C_{20} and C_{24} n-alkanes were co-injected to monitor analytical precision, which was 0.2 per mill and 0.3 per mill, respectively for both standards. GDGT fractions were analysed according to Schouten et al. (2007a), Kuypers et al. (2002). Single ion monitoring of the $[M + H]^+$ ions (dwell time, 234 ms) was used to quantify the tetraether lipids with 1-4 cyclopentyl rings and calculate the TEX_{86} values. These values were converted to sea surface temperature (SST) according to the equation:

$$TEX_{86} = 0.027 * SST - 0.016 \qquad (1)$$

Equation (1) is updated from Schouten et al. (2003), and is based on correlating TEX_{86} values of 43 core-top sediments with annual mean SSTs ranging between 20 and 28 °C. Replicate analysis has shown that the error in TEX_{86} values is 0.01 or 0.6 °C. A record of 61 samples for the MP and likewise for the VB was established achieving striking results.

3. Study Area

During the lower Albian (112 ma bp), both investigated sites were located in the palaeotropical climate belt (Fig. 13) but differ in their chemical composition due to regional varieties in palaeoclimatology and palaeogeography.

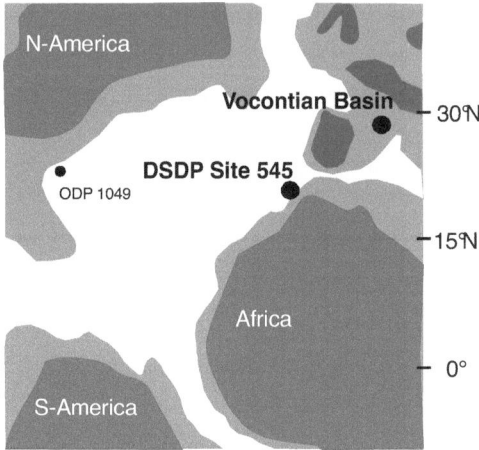

Fig. 13: Location of the investigated OAE 1b sites (Erbacher et al. 2001; Herrle et al. 2004) superimposed on the palaeogeographic map of the mid-Cretaceous at ~110 Ma (after Scotese 2001).

This instance was chosed to compare both OAEs in differing regional settings but in an superordinate climate system where the main processes are connected to each other. In addition, this study presents information dealing with this differences showing up their effects on the land-ocean system. In the following, each site is discussed in detail before a comparison in the Synthesis is strived. Although the aim was to achieve a similar geochemical data record by using the same technical approach, the analysed material is very different in its chemical composition so some detours are attached. Nevertheless, these sites were chosen because an extensive sample record was already available analysed by numerous previous workers (e.g. Bréhéret 1985, 1994; Erbacher et al. 1999, 2001; Leckie et al. 2002; Herrle 2002; Hofman et al. 2001, 2008; Kuypers et al. 2002; Friedrich et al. 2005; Wagner et al. 2007 etc.). The aim was to complete these records with a multiplicity of analyses, trying to solve the circumstances leading to the formation of this black shale.

3.1. General remarks

Each study section comprises a multitude of results, voluminous in detail yielding to predictions about palaeoenvironmental situations during the Lower Albian. Aim of this study was furthermore to compare these two sites and point out regional differences as well as similarities reflecting palaeoclimate and –oceanography. Presented results should advance the still not clearly known circumstances that led to the deposition of OAE 1b with a newly developed bulked data record representing the most possible highest resolution presented in 1 cm increments in the order of 700-800 years per sample. This entire work is divided into two self contained parts, describing each sampling site on the whole including introduction, results and the discussion. In the following Synthesis chapter, all results and presumptions from both sites are collected and compared trying to solve questions about black shale formation and the resulting perturbation on the global climate system in regard to regional distinctions. Furthermore, questioning the black shale event as a regional or rather a global event will be discussed.

A short overlook over previous studies will be presented in this chapter. Numerous scientist have investigated in either one of the sampling sites or in both. This fact was helpful, adding expressive information to the here presented record. A complete data record is established, including various geoscientific-measurement methods presenting a detailed overview on several circumstances during Albian "Greenhouse" conditions.

3.1.1. DSDP Site 545, Mazagan Plateau

The Mazagan Plateau (MP) was first described in detail when it was drilled in 1981 by the Shipboard Scientific Party of Leg 79 (Fig. 14). On a total of 300 samples (from Site 544-547) LECO and Rock-Eval analyses were obtained on board. Seventeen samples were

STUDY AREA

selected to be extracted with chloroform and ten of these samples were analysed by gas chromatography of saturate hydrocarbons. Elemental analysis were performed on the same ten samples (Deroo et al. 1984). A negative correlation between carbonate content and organic carbon content was observed for the Aptian and Albian deposits of Hole 545. The nature of the organic matter belong either to the marine organic matter that was altered during deposition or to an alteration of well-preserved and highly altered marine organic matter. Non marine matter is only detected at the base of the late Aptian sequence. The sedimentological describtion and the classification of the planktonic foraminiferal zone was accomplished by Leckie (1984, 2002) . Herrle (2002) commenced a high resolution study of calcareous nannofossils of the OAE 1b succession. For OAE 1b, Erbacher et al. (1996b) demonstrated its correlation to transgressive periods resulting from increased oceanic productivity. Anoxic meso- to eutrophic conditions appear to have been very strong during deposition of the Niveau Paquier indicated by a lack of benthic foraminifera within the black shales and relatively slow repopulation.

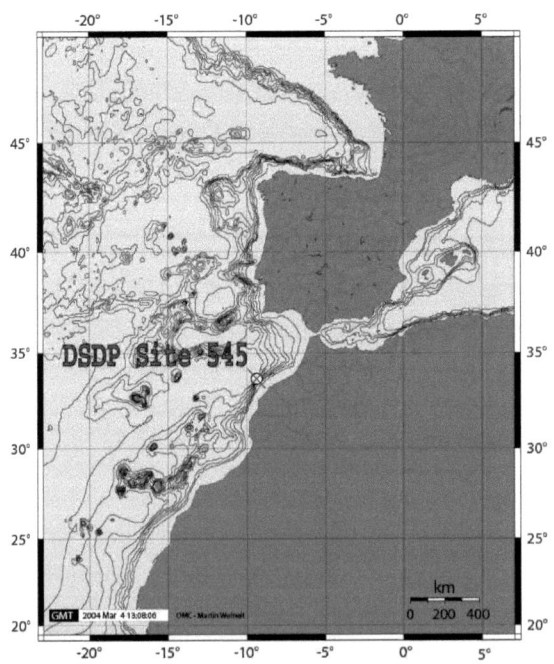

Fig. 14: Location of DSDP Site 545 at the Mazagan Escarpment offshore Morocco (map created with OMC 2004).

Stable carbon isotope ratios from the MP were already analysed receiving a time resolution of ~ 1 to ~ 55 kyr (Herrle 2002; Herrle et al. 2003) but only on a total of 65 samples. The critical interval including the black shale (core 42, about 9 metres) was re-sampled for the here presented study achieving a total of 300 samples. Leckie (1984) calculated a mean sedimentation rate of ~ 2 cm/kyr. Stable carbon and oxygen isotope data were measured on mid-Cretaceous sediments from ODP Site 1049 in the western subtropical Atlantic, off Florida, including a 46 cm thick early Albian OAE 1b (Erbacher et al. 2001). Erbacher et al. calculated an age model, based on sedimentation rates, with an estimated duration of ~ 46 kyr for OAE 1b formation in the Atlantic (Erbacher et al. 2001; Herrle 2002). Leckie postulated significant upwelling and associated fertile surface waters from Site 545 off central Morocco during latest Aptian through middle Albian time (Leckie 1984). His conclusion was supported by the accumulation of siliceous skeletons (radiolarians and sponge spicules) and their presence as the dominant biotic elements.

Thereupon an intensified oxygen minimum zone during the latest Aptian and early Albian is probably responsible for the disappearance of several planktonic taxa. The vertical expansion of the oxygen minimum zone peaked during the middle Albian (Cores 545-43 to 545-41) as recorded by the deposition of black shales. Data records from the Blake Nose and the Mazagan Plateau show a synchronous drop in $\delta^{18}O$ and $\delta^{13}C$ values suggesting an input of isotopically light carbon into the oceans and a rising sea surface temperature of 4-5 °C (Jenkyns 2003; Leckie et al. 2002). This signal is seen regionally in bulk carbonates from the Atlantic and southern France. In contrast to the Vocontian Basin, were the sediments may carry a variable burial-diagenetic overprint (e.g. Weissert & Bréhéret 1991) the isotopic data from the Atlantic are deemed the more reliable (Jenkyns 2003).

Herrle and Herrle et al. (2002, 2003) developed a cyclostratigraphic time-scale based on the recognition of eccentricity, obliquity and precession cycles in the sedimentary record. Their results indicate that the temperature rise took place over a period of 40 ka, with the initial trigger being much more rapid than this. An explanation model by Erbacher et al. (2001) suppose increased carbon burial related to water mass stratification caused by an increase in run-off into semi-restricted basins like the Tethys and the proto-Atlantic. They drew a parallel to the formation of Quaternary Mediterranean sapropels. Hofmann et al. (1999) drew attention to complex cycle patterns in the late Albian North Atlantic suggesting the interaction of a coupled climate-ocean system. On selected DSDP Sites near the continental margins (369 and 603), climate cycles, reflected by the oscillating clastic supply, are linked to ocean fertility which was highest during humid periods. The authors drew a connection between orbital forcing and delicately balanced oceanographic changes in the carbonate-carbon budget. Hofmann et al. (2001) suggest higher amounts of precipitation during the Albian for northern Africa and parts of southern Europe, accompanied by latitudinal shifts in the major climate belts. Also Herrle (2003) suggest that a monsoonal climate was dominant and that fluctuating surface-water fertility conditions was the main favouring factor for a black shale deposition.

3.1.2. Vocontian Basin, SE-France

The Lower Albian black shales were initially described from southeastern France, in the Vocontian Basin, with coeval correlatives in Germany and Austria and have nowadays been traced through the Atlantic-Tethyan realm (e.g. Bréhéret 1997, Bralower et al. 1999; Leckie et al. 2002; Tsikos et al. 2004). The sediment successions of the Vocontian Basin were studied in different sections inter alia the L´Arboudeysse section comprising the black shale Niveau Paquier (OAE 1b). The l´Arboudeysse outcrop section is situated in the northern part of the VB (Fig. 15), 5 km east of the village of Rosans (topographic map 25 Rosans, No. 3239 Ouest, Série Bleue), and exhibits an 1.63 m thick OAE 1b bed (Friedrich et al. 2005).

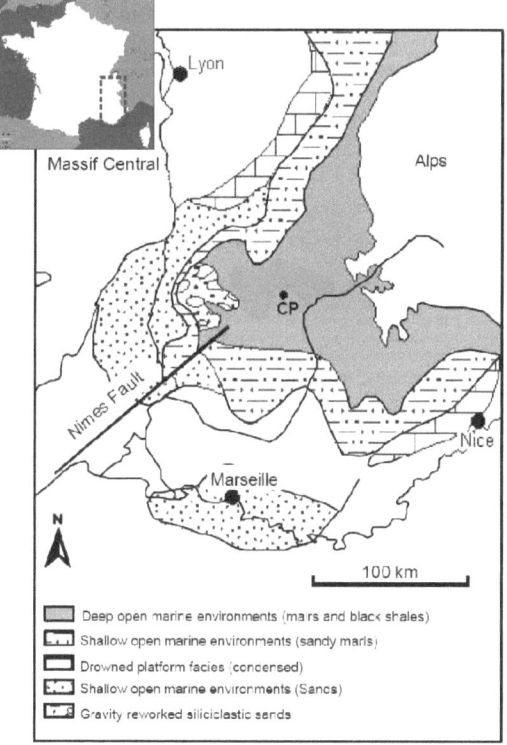

Fig. 15: Palaeogeographic map of the Vocontian Basin during the Late-Aptian and Albian (after Arnaud & Lemoine 1993) showing the location of the studied section Col de Palluel (CP) with l´Arboudeysse.

STUDY AREA

On the 14 meter long record the black shale event begins at meter 8.40 and ends at meter 9.95. It was first recognised and described by Bréhéret (1985). Excellent stratigraphic classification of the Vocontian Basin (VB) have resulted from intense studies (e.g. Moullade 1966; Bréhéret 1988) and were expanded by first microfaunal content studies (Erbacher et al. 1998). Erbacher et al. (1998) used Q-mode principal component analyses on a sample selection for testing the composition of benthic foraminiferal assemblages. High-resolution study of calcareous nannofossils, palynomorphs and benthic foraminifera as well as oxygen isotope records were conducted by Herrle (2002 and Herrle et al. 2003). Friedrich et al. (2005) analysed benthic foraminiferal repopulation events and their forcing mechanisms for the VB and the MP. Herrle & Mutterlose (2003) show a high abundance of *R. parvidentatum* during the uppermost Aptian-Lower Albian, indicating a boreal influx into the Tethyan Realm and reflecting a generally cool climate. These climate conditions were accompanied by a general rise in sea-level.

The results show short-term ventilation of intermediate to deep waters, a supraregional distribution for OAE 1b, a precessional signal within the Milankovitch band and similarities to sapropels (age: Pliocene-Holocene) of the Mediterranean Sea (also discussed by Erbacher et al. 2001). Based on biostratigraphy, sedimentological parameters and nannoplankton associations, an age model was established based on the recognition of eccentricity, obliquity and precession cycles in the sedimentary section. The distribution and composition of calcareous nannoplankton is influenced most dominant by the nutrient availability and temperature in the surface waters. By performing Principal Component Analyses (R-PCA), these parameters were quantified into nutrient and temperature indices (NI; TI). The complete interval analysed by Herrle (2002) comprises a duration of ~ 374 kyr and the OAE 1b formation occurred within ~ 44 kyr. A first $\delta^{13}C$ record (established by Herrle et al. 2003) was proposed as a standard reference section for stratigraphical correlation of the Aptian to Lower Albian. Herrle (2003) suggest that monsoonal climate and strongly fluctuating surface water-fertility were dominant conditions during black shale deposition.

First TOC, $CaCO_3$ and Rock Eval data sets were established by Bréhéret (1994a) presenting insights of the organic matter origin. Biomarkers detected in OAE 1b from ODP Site 1049 in the western North-Atlantic and in the VB have shown the extensive occurrence of archaea bacteria, contributing up to 40-80% to the bulk sedimentary organic matter (Vink et al. 1998; Kuypers et al. 2001, 2002). This very high contribution of nonthermophilic, marine archaeal origin is unique and demonstrates clearly that the origin of marine organic matter (archaea versus phytoplankton) and the causes for $_{13}$C-enrichment may be completely different despite the apparent similarities between marine black shales of Mesozoic age (Kuypers et al. 2001).

Herrle et al. (2003) proposed that monsoonal activity, resulting from precessional forcing and modulated by eccentricity-driven temperature changes, represents the driving factor of OAE 1b formation in the VB as well as increased productivity. A strong monsoonal activity is characterised by humid conditions and strong winds. Changes in the position of the nutricline were probably controlled by changes in monsoonal activity, based on the calcareous nannofossil approach (Herrle 2003). They link the supraregional occurrence of the black shale to enhanced preservation of organic matter due to a reduction of deep water formation in the low latitudes under extremely warm and humid conditions (referring to Erbacher et al. 2001).

4. DSDP Site 545, Mazagan Plateau

4.1. Palaeogeography and Geology

DSDP 545 was drilled in 3150 m water depth at the foot of the steeply sloping Mazagan Plateau Escarpment offshore Morocco, 200 km SW of Casablanca (Fig. 14; Winterer and Hinz 1984). The Moroccan Atlantic continental margin constitutes an example of passive margin due to the Atlantic opening between African and North American Plates (Mhammdi & Bobier 2001). The structural relief of the Mazagan Escarpment was created in the Middle Jurassic when the earliest seafloor spreading began in the central Atlantic (Winterer & Hinz 1984). The stratigraphic development of this passive continental margin was driven by subsidence and tectonic processes. Nowadays, Triassic to Cenozoic sediments superpose the crystalline basement. In the Early Cretaceous, Africa was one of three large continental blocks with shallow seas (Hay et al. 1999). Palaeogeographic reconstructions suggest that the Early Cretaceous Atlantic became a site of organic carbon burial because of isolation and restrictions due to smaller plates and other shallow blocks. During the lower Albian, Site 545 was located at about 2000 m palaeowater depth mainly well above the CCD (Winterer & Hinz 1984) at a palaeolatitude nearby 20°N (Hay 1999). After Voigt et al. (1999), Site 545 was located in the tropical climate belt (between 10°N - 30°N palaeolatitude), characterised by the g reatest deficit of precipitation minus evaporation. The excess tropical heat may have contributed to the intensified evaporation and aridity that is predicted between 10°N - 25°N p alaeolatitudes (Ufnar et al. 2004).

Sediments documented for the Albian consist mainly of green nannofossil claystones composed chiefly of illite and mixed-layer clays (Shipboard Scientific Party 1984). They are about 110 m thick and a general decrease in planktonic foraminiferal diversity leads to the suggestion of an expanded oxygen-minimum zone for this time period (Leckie 1984). Investigations were focused on Core 42 which contains the OAE 1b corresponding interval of the *Hedbergella planispira* planktic foraminiferal zone and the Prediscosphera columnata NV8B nannoplankton subzone (Fig. 16, e.g. Leckie 1984, Leckie et al. 2002 and references therein, Herrle 2002). The OAE 1b interval comprises an 80 cm thick black shale extending from 389.85-390.65 metres below sea floor (mbsf; Leckie et al. 2002). The black shale horizon is generally faintly laminated. The top 15 cm are mottled burrows which are up to 0.5 cm in diameter. The OAE 1b black shale is located in a succession of bioturbated, light grey-weathering, nannofossil-rich marls.

Site 545 is specifically suitable for a detailed climate study because stratigraphic control and sedimentary boundary conditions are reasonably well constrained (Hofmann et al. 2008; Leckie et al. 2002; Herrle 2002). Leckie et al. (2002) described OAE 1b as a watershed event in the evolution of planktonic foraminifera and as a marker for the beginning of the end of widespread black shale deposition and the initiation of chalk deposition. During latest Aptian through middle Albian time, site 545 was characterised by high productivity due to enhanced upwelling and associated fertile surface waters (Leckie 1984; Poulsen et al. 1998). Samples were taken from core 42, Section 1 to 4, leading to a sample record from 6 metres, 388,5-394,5 depth in meter below sea floor. The black shale OAE 1b is recorded as a 80 cm thick layer in 389,85-390,65 mbsf with a gradational contact of the black shale and the nannofossil-rich claystone (own observations; Shipboard scientific Party 1984). Hard grounds, interruptions, or sedimentary gaps across the studied interval are not observed (Fig. 17).

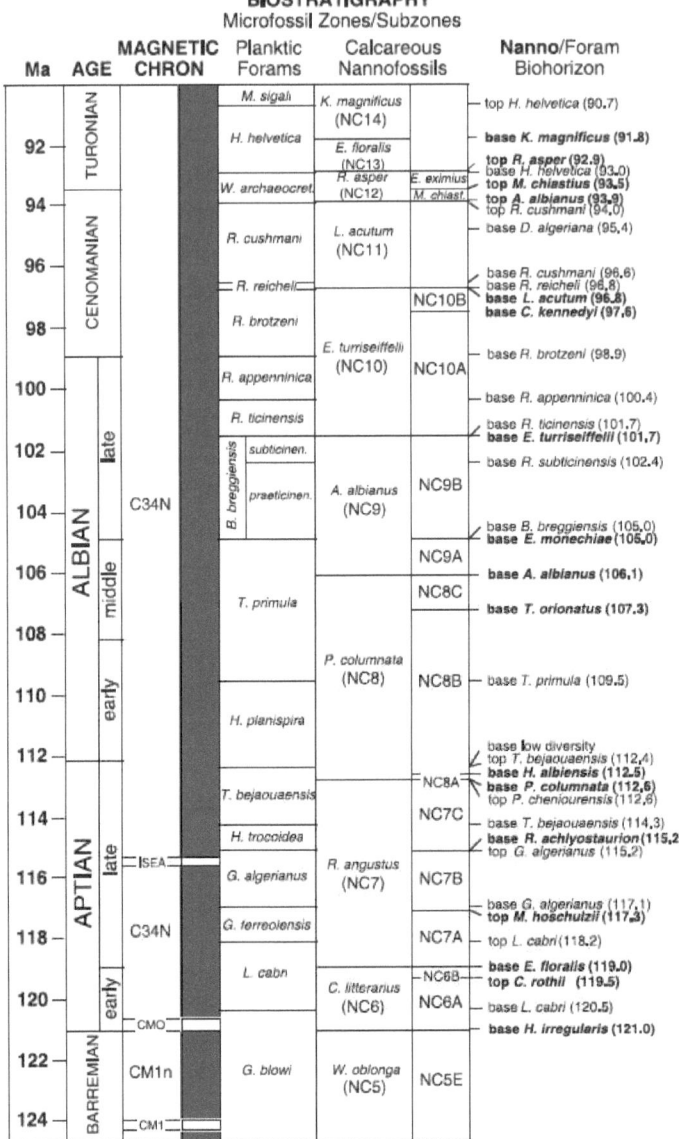

Fig. 16: Integrated calcareous plankton biostratigraphy (Leckie et al. 2002).

Fig. 17: Photos taken from the analysed core 42, DSDP Site 545, archive and working half. A) The beginning of OAE 1b at 390,65 mbsf and B) the end of OAE 1b sedimentation (at 389,85 mbsf) show both no interruptions or gaps and reveal a gradational transition between black shale and claystone.

4.2. Results

In respective sub-chapter the results will be described in detail and shortly discussed. Trying to deal with a huge amount of achieved data, the focus of the introducing discussion is to demonstrate the objectives of the measurements and which conclusions are feasible concerning the results. The main data set will be presented in either XY ore triangle plots against depth. Due to the technical approach, presentation of the organic matter will include gas chromatograms and HI/OI diagrams. The ensuing discussion chapter (see 4.3.) highlights a summing up picture of the data record, putting together all results and presenting them in a threefold system before-, during- and after the OAE 1b event. Most of the here presented results are published in: Hofmann et al. 2008, Wagner et al. 2007 and Wagner et al. 2008.

4.2.1. Stable carbon isotopes

The results confirm, that the OAE 1b interval can be defined by the bulk carbonate and bulk organic carbon $\delta^{13}C$ isotope records (Wagner et al. 2007, 2008). The black shale interval starts and ends simultaneously with the already mentioned negative isotope excursion. Pre-OAE $\delta^{13}C_{carb}$ values are constant at ~2 ‰ without showing any repeating patterns and rapidly shift by ~1.5 ‰ to more negative values at the base of OAE 1b (390.65-390.70 mbsf; Fig. 18).

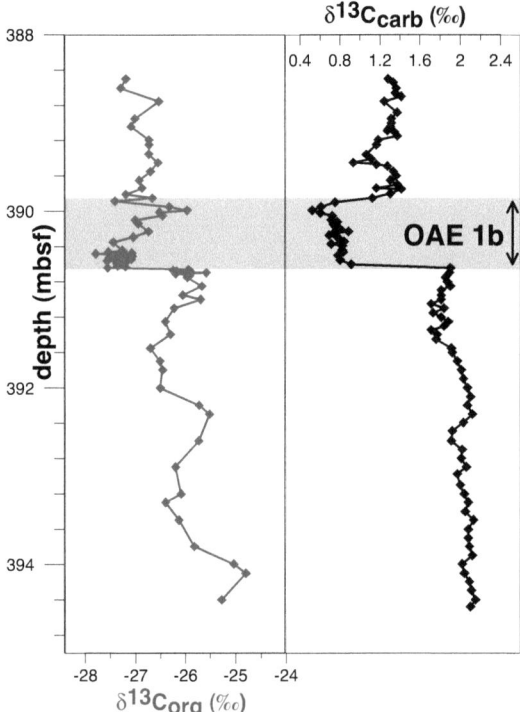

Fig. 18: Bulk stable carbon isotopes from the MP. The grey shaded interval marks the duration of OAE 1b. A lower resolution bulk carbonate $\delta^{13}C$- record (provided by Herrle 2002) was expanded to achieve high resolution for this study. Note the abrupt shift at the black shale onset.

This shift is very pronounced and abrupt occurring within a few centimeter, which means presenting only a few thousand years. But the surprising fact is that the magnitude of the shift remains constantly negative over several tenthousand years and is reaching even more light values throughout the event. Most of the Cretaceous black shales are initiated by a shift in the carbon isotopes but rather as a short pulse, normally a negative shift followed by a positive excursion reflecting the duration of the event (e.g. OAE 1a and OAE 2). The black shale event at the MP covers 80 cm and is defined by the carbon isotope excursions from 390.65-389.85 mbsf. $\delta^{13}C_{carb}$ remain at constant negative levels throughout the first part of the OAE, then slightly drop another ~0.25 ‰, before they gradually recover to achieve post-OAE levels of ~1.5 ‰ slightly exceeding those recorded before the event (Fig. 18). The termination of the OAE is characterised by a gradual return to isotopic values in the order of 1.2 ‰ (Hofmann et al. 2008; Wagner et al. 2007, 2008; Herrle 2002). After the event isotopic values still seem disturbed but within a short recovery phase steady state conditions are reached again.

The ^{13}C record of OC ($\delta^{13}C_{org}$) parallels that of $\delta^{13}C_{carb}$ although demostrating some different trends. During the pre-OAE, more variable 0.5-1 ‰ swings are obvious, presenting a somewhat cyclic pattern fluctuating from –25 ‰ up to nearly –27 ‰. The post-OAE, records more clearly the state of disturbance compared to the pre-OAE state, swinging apparently undefined between –26 ‰ and –27 ‰. The onset of the black shale is marked by a negative shift of ~1.5 ‰ followed by a graduall return to less negative values up to -26 ‰ in the upper part of the OAE, and a general decrease towards the top of OAE. Notably, $\delta^{13}C_{carb}$ and $\delta^{13}C_{org}$ reveal a parallel negative shift of the same magnitude at the base of the OAE.

The record suggests a constant supply of ^{13}C-depleted CO_2 during the entire black shale interval because the abrupt negative shift remains constant and excludes the possibility of a single degassing event, which than would have been constantly exhausted.

4.2.2. Organic composition

The organic matter content pre- and post dating OAE 1b at the MP fluctuates between 1 and 2.5 % total organic carbon (TOC; Fig. 19). Directly prior to the onset of OAE 1b, TOC values increase by approximately 1 % from 1.5 to 2.5 %. The OAE 1b interval is characterised by a gradual increase in TOC to almost 5 % (at 389.9 mbsf) followed by a rapid return to TOC values comparable to pre-OAE concentrations in the top ~10 cm. The TOC content of the OAE 1b interval is significantly lower than at other OAE 1b Sites of the North Atlantic, e.g. ODP Site 1049, where up to 13 % TOC are reached (Hofmann et al. 2008; Barker et al. 2001; Kuypers et al. 2001). The drop in TOC values in the top of the black shale is sharp but the onset of the OAE is accompanied by a slight increase. The very abrupt shift in the carbon isotopes, marking the beginning of the black shale, is confirmed by the organic composition.

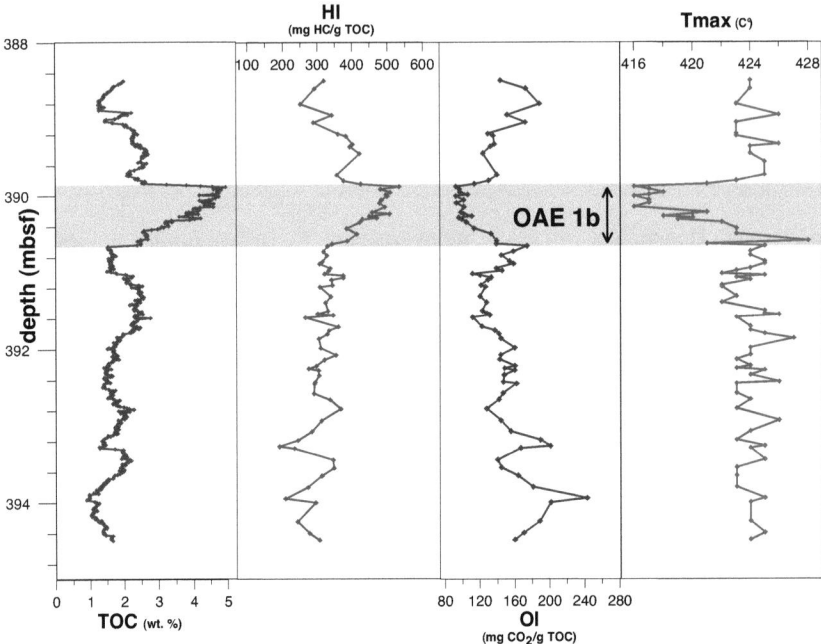

Fig. 19: Geochemical record of the complete analysed section of Site 545. Organic composition: TOC, HI, OI and Tmax.

Hydrogen indices (HI), an indicator of the hydrogen-richness of the bulk organic matter (OM) and at high values commonly associated with mainly algael and bacterial biomass (Hofmann et al. 2008), follow changes in TOC and range from 190 to high 530 mg HC/g TOC values (Fig. 19). Oxygen indices (OI), a measure of oxygen richness of the bulk OM with high values supporting terrestrial OM but alternatively an indicator of oxygenation at times of OM preservation, range from 90 to 240 mg CO_2/g TOC (Fig. 19) In general HI and

OI values show an overall inverse behavior for the entire interval of investigation. HI values > 450 HI mg HC/g TOC and OI values < 150 mg CO_2/g TOC are restricted to the OAE 1b interval. HI values show a gradual increase to higher values over the OAE 1b interval with a sudden decrease at the top of OAE 1b and thus a pattern similar to TOC content. Plotting the data in an *van Krevelen* typ diagram (Fig. 20) the values indicate that the organic material has a mixed kerogen character of type II/III (a mixture of marine and terrestrial material) but mainly type II kerogen (Hydrogen indices ~300 mg HC/g TOC with values exceeding 500 mg HC/g TOC in the black shale; Wagner et al. 2008). Pyrolysis temperatures (Tmax, Fig. 19) below 425 °C indicate that the OC is immature.

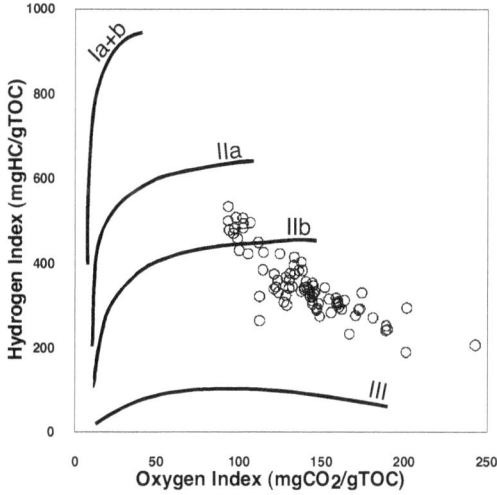

Fig. 20: Projection of the Rock Eval data in a modified van Krevelen diagram (after Tyson 1995).

Isotopically heavy archaeal-derived OM, characteristic for non upwelling OAE 1b sites in the western Tethys and at Blake Nose (Kuypers et al. 2001; Tsikos et al. 2004), is not present at the MP, probably owing to overall high nutrient level of surface waters off northwest Africa at that time (Leckie 1984). This is also supported by the abundance of sterol ethers at Site 545 that have also been identified in sediments deposited below modern upwelling regions (Schouten et al. 2005). Probably the high amounts of phytoplankton off northwest Africa, marking high productivity areas were dominant and obviated the invasion of archaea. The carbonate carbon record ($CaCO_3$) varies inversely to TOC. Concentrations vary between 22 and 52 % $CaCO_3$ (Fig. 21). The pre-OAE interval is characterised by cyclic carbonate fluctuations. With the onset of OAE 1b the $CaCO_3$

content gradually decreases to a minimum of 22 % and progressively increase to ~40 % $CaCO_3$ at the top of the OAE interval. The base of the post OAE interval displays faily constant carbonate content values around 45 % before the renewed onset of a more cyclic fluctuation pattern towards the top of the section is re-established. The observed carbonate content fluctuations correlate well with the total abundance of calcareous nannofossils reported by Herrle (2002) for this interval, suggesting that $CaCO_3$ content is mainly controlled by calcareous nannofossil abundance. Apparently, in contrast to TOC the carbonate trend is hardly interrupted at the critical interval marking the onset of the black shale event. Carbonate reaches lowest values during the middle segment of the event (around 390.3 mbsf) with a gradual return to a higher $CaCO_3$ content at the top of OAE 1b. In fact, values are not extremely different during the event than they were prior to it and the rhythmic pattern remains stable. However, after the event $CaCO_3$ values reveal constant and a returning to a more fluctuating pattern is only in the upper part of the record conceivable (from 388.9 mbsf). With the exception of three distinct total sulphur (S_{tot}) peaks, the sulphur content of the entire section lies in the order of 0.8 % (Fig. 21).

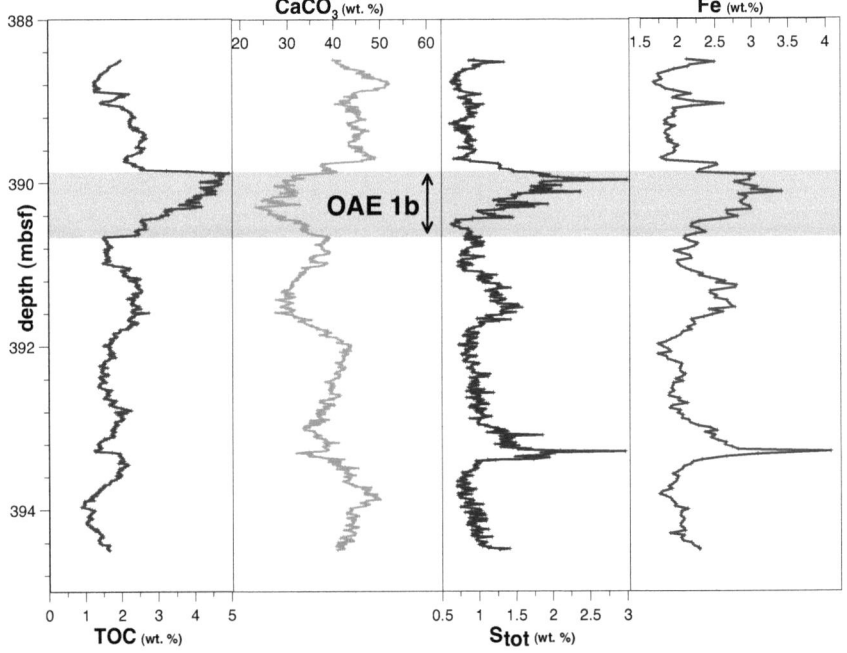

Fig. 21: Geochemical record of the total organic carbon- (TOC), the carbonate carbon- ($CaCO_3$), the total sulphur- (S_{tot}) and iron (Fe) content.

The maximum sulphur concentration is reached towards the top of OAE 1b and corresponds to maximum TOC concentrations in this interval. Higher sulphur concentrations are generally recorded during cooling episodes following warmer episodes as indicated by the TEX_{86} derived sea surface temperatures (SST; next chapter). Sulphur and Fe- concentrations correlate, suggesting iron limitation for the fixation of sulphur in the sediments (Berner & Raiswell 1985). This interpretation is supported by the observed TOC-Fe-S_{tot} relationship. Most of the studied samples plot in the ternary TOC-Fe-S_{tot} diagram along a line extending from the TOC corner to the Fe-S_{tot} line which is typical for iron-limited systems (Dean & Arthur 1989; Emeis & Morse 1993; Arthur & Sagemann 1994; Hofmann et al. 2000).

4.2.3. Biomarker

Studies in modern oceans have shown that TEX_{86} mostly records sea surface temperatures (SST; Schouten et al. 2002; Kim et al. 2008) though in some upwelling areas, e.g. Santa Barbara Basin (Huguet et al. 2007), the TEX_{86} tends to underestimate actual SSTs (Kim et al. 2008). SST estimates based on TEX_{86} calculations for the MP range from ~29°C to 33°C (Fig. 22) and are in the s ame range as SST estimates for the Cretaceous oceans based on modeling and $\delta^{18}O$ estimates from pristinely preserved planktic foraminifera (Poulsen et al. 2001; Wilson & Norris 2001; Norris et al. 2002; Huber et al. 2002; Schouten et al. 2003; Steuber et al. 2005; Bice et al. 2006; Forster et al. 2007; Bornemann et al. 2008). Therefore, the TEX_{86} data is interpreted as SSTs although actual Albian SSTs may have been even somewhat higher. The SST profile reveals an overall warming trend by approximately 1°C over the investi gated interval with superimposed cyclic fluctuations. Moderate warm/cold cycles with a temperature range of 1.5°C are clearly recognised in the lower, pre-OAE section of the record with temperature values of ~29°C. This general trend in SST is disrupted durin g OAE 1b. The onset of the event is marked by a rapid temperature increase by 3.5°C at its base which is more than twice as high as the observed temperature fluctuation in the pre-OAE section. The rapid increase is concomitant with the negative shift in the $\delta^{13}C$ isotopes. Sea surface temperatures persist high, between 32°C and 33°C, across the event and a variable return to ~30°C with superimposed warm/cold cycles following the termination of OAE 1b (Fig. 22).

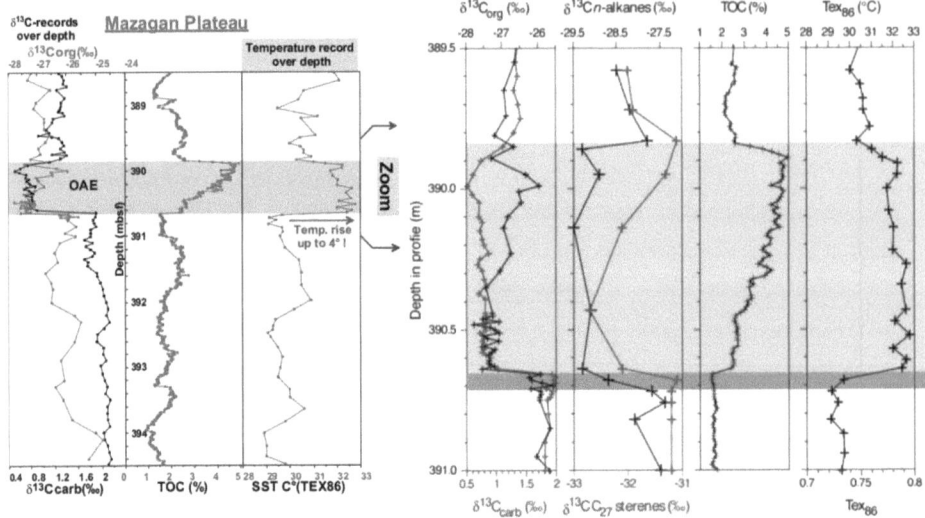

Fig. 22: <u>On the left</u>: Geochemical record of carbon isotopes, temperature record and organic composition at the Mazagan Plateau. The SST is based on the TEX$_{86}$. The TEX$_{86}$ (TetraEther indeX of tetraethers consisting of 86 carbon atoms) is determined from crenarchaeota biomarker. <u>On the right</u>: Focused on the black shale interval supplemented by compound-specific stable carbon isotope data from marine C$_{27}$-sterenes and terrestrial long-chain n-alkanes (after Wagner et al. 2008).

Beside the already discussed δ^{13}C carbonate and organic carbon isotopes, furthermore the compound-specific carbon isotopes from terrestrial plant leaf waxes (C$_{27}$, C$_{29}$, C$_{31}$ n-alkanes) and marine algael C$_{27}$-sterenes were measured. The Carbon Preference Index (CPI), a measure of thermal maturity and dominance of terrigenous organic matter using nC$_{24}$ to nC$_{32}$ n-alkanes, was calculated to support the discussion on sources of organic matter (Wagner et al. 2008). The CPI of the terrestrial plant leaf waxes ranges from 2.5 before and after the excursion to up to 7 within the TOC maximum. The compound-specific carbon isotopes follow bulk ^{13}C (Fig. 22), i.e. they are ca. -28.5 ‰ for n-alkanes and -31 ‰ for C$_{27}$-sterenes before the OAE, then rapidly shift by 1-2 ‰ to more negative values at the base of the OAE, and gradually return close to their pre-OAE values.

Just below the base of the event (at 390.75 mbsf), the δ^{13}C values of n-alkanes start to decline to lighter values while δ^{13}C of sterenes do so ~7 cm higher in section (at 390.68

mbsf), concomitant with the onset of OAE 1b. As for the bulk isotopes, both molecular records reveal slightly different trends, i.e. the n-alkanes tend to follow $\delta^{13}C_{carb}$ except for the base of the OAE whereas the C_{27}-sterenes are similar to $\delta^{13}C_{org}$. As the isotopic composition of terrestrial material primarily reflects the isotopic composition of atmospheric CO_2, the sudden negative spike in ^{13}C of n-alkanes must be due to a change in the ^{13}C of atmospheric CO_2. This observation indicates that ambient atmospheric carbon reservoirs and SST changed almost instantaneously. It is striking that the onset of the negative shift in the n-alkanes slightly precedes that of the marine C_{27}-sterenes and both bulk ^{13}C records, strongly suggesting that ^{13}C-depleted carbon was abruptly released into the atmosphere.

Contrary to the OAE 1b occurrence at the Vocontian Basin the isoprenoid biomarker parameter Lycopane (2,6,10,14,19,23,27,31 –octamethyl dotriacontane) is not apparent at Site 545. Lycopane is an example of a head-to-head isoprenoid which are specific markers of archael input to sediments. It originates from an unsaturated to polyunsaturated C_{40}-isoprenoid. These alkenes can be preserved as sulphur-bound species under anoxic marine conditions (de Leeuw & Sinninghe Damsté 1990). Lycopane is widespread and is particularly abundant in lacustrine sediments that were deposited under anoxic conditions. It is fairly common in modern marine sediments (e.g. the Peru upwelling zone) as well as marine source rocks (e.g. the Monterey Formation; Schouten et al. 1997a). As it is better preserved in anoxic conditions it is attributed as a parameter reflecting oxic to anoxic conditions (Peters et al. 2005). Lycopane typically elutes near nC_{34}-nC_{35} and is measured by GCMS using various mass fragments characteristic of the tail-to-tail isoprenoid linkage. In the here presented study, there was no advice for the existence of Lycopane at the Mazagan Plateau.

Another biomarker parameter, although specificity is rather low, is the Pristane/Phytane ratio. Here, the biological origin is presumably archael and environmental conditions are thought to present anoxic conditions or high salinity. Problems in significance are due to interference by thermal maturity and source input. Pristane (C_{19}) and Phytane (C_{20}) are commonly derived from the phytyl side chain of chorophyll a in phototrophic organisms and bacteriochlorophyll a and b in purple sulphur bacteria (Peters et al. 2005). The Pristane/Phytane ratio varies between 0,2-narrow 2% with a mean average of 0,8% (Pr/Ph-ratio normalized on TOC), indicating anoxic or hypersaline conditions without a specific prediction about palaeoenvironmental conditions. In general, their abundance is very low and not identified in the specific carbon isotope compound analyses.

4.1. Inorganic composition

The presented results are divided into the clastic flux composition and the redox-sensitive elements. Beginning with the clastic flux composition quartz and clay minerals are considered as essentially detrital and wind –blown fraction during the Cretaceous. Maximum dust input remained in palaeolatitudes 20-30°N since early Cretaceous. The presented data show compositional variations in the siliciclastic fraction across the studied section, reconstructed from aluminium normalised metal ratios (see also Hofmann et al. 2008). Si/Al and Zr/Al are thought to reflect fluctuations in grain-size and the K/Al ratio, or illite vs. smectite and kaolinite, is indicative for compositional variations in the clay mineral assemblage (Fig. 23). The variations in Mg/Al ratios is used as an approximation of the palygorskite/smectite+kaolinite ratio.

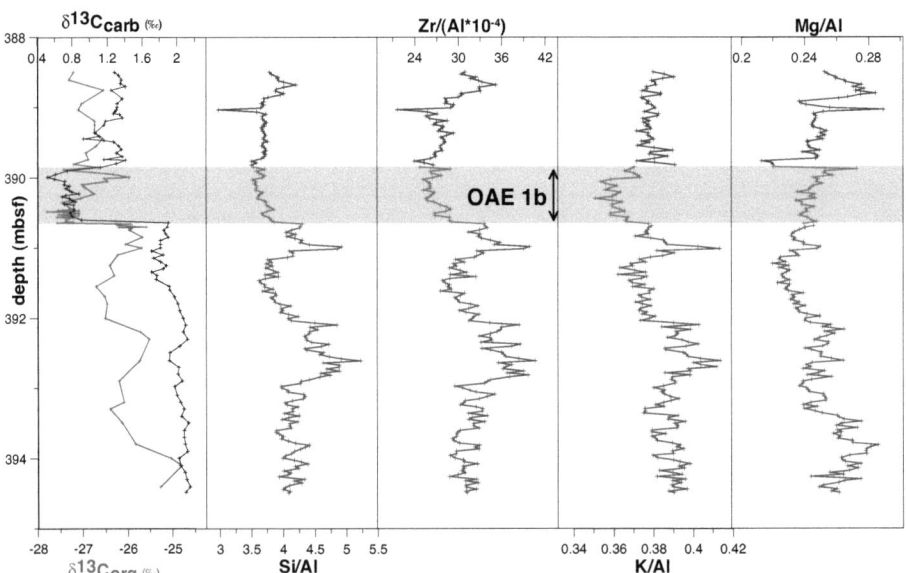

Fig. 23: Clastic flux composition at the MP. Si/Al and Zr/Al reflect fluctuations in grain-size, K/Al compositional variations in clay minerals and the Mg/Al as approximation of the palygorskite + kaolinite ratio.

Records of Si/Al and Zr/Al show cyclic, low amplitude variations in the pre OAE part of the section. High Si/Al ratios of 3,5-5 support the presence of elevated amounts of quartz, considering that Si/Al ratios of the widespread clay minerals smectite and illite range between 2 and 3 (Weaver and Pollard 1975). Si and Zr and, to a smaller degree, Si and Ti correlate significantly (Fig. 24) which indicates a common source for Si and Zr. The elements Zr and Ti are typically concentrated in the heavy mineral fraction (Taylor and McLennan 1985) as zircon, rutile or other Ti-bearing phases. The correlation of Si with Ti and Zr supports a common terrestrial source for the three elements. X-ray diffraction data from the clay mineral fraction revealed that the dominant clay minerals are smectite (Al-rich) and illite (K-rich), followed by palygorskite (Mg-rich) and irregular mixed-layer clays and support the suggested predominantly detrital origin for the clay assemblages (Chamley & Debrabant 1984). Detrital illite results from physical weathering whereas smectite is a common product due to chemical weathering from altered volcanic rocks or kaolinite, illite and mixed layers. Physical weathering is dominant in regions with highly contrasting climate zones, preferable hot, cold and dry.

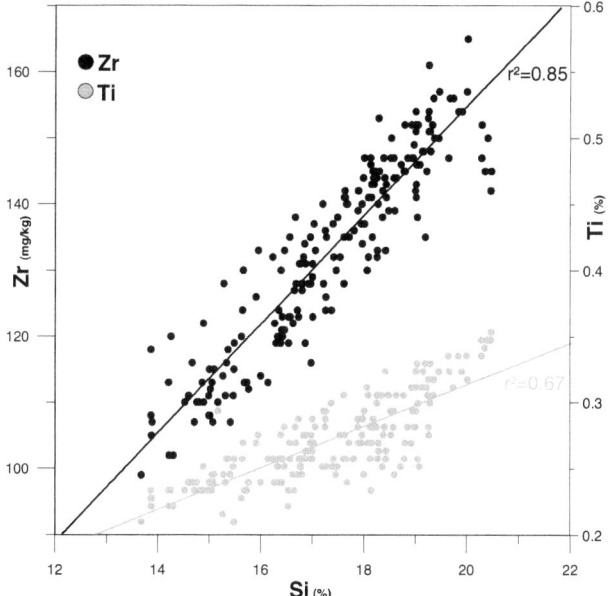

Fig. 24: Correlation of Zr, Ti, and Si. The good correlation suggests that all elements are delivered to DSDP Site 545 as part of the siliciclastic flux.

Weathering of aeolian silt and sand issued from W-Africa is characterised by the preservation of detrital illite, kaolinite and palygorskite aggregates (Chamley 1989). At Site 545, kaolinite and chlorite only occur in lower concentrations. K/Al ratios may hence serve as a proxy for the relative abundance of illite over smectite.

Illite is generally formed during sediment diagenesis and is often recycled as detrital sediment components whereas smectitic clays are found in hydromorphic soils or form during hydrolyzing climate conditions and thus represent moist climate conditions (Chamley 1989; Weaver 1989; and references therein). Illite abundance as indicated by relatively high K/Al ratios follows the pattern of other detrital sediment components, e.g. zirconium and quartz abundance (as revealed by Zr/Al and Si/Al ratios; Fig 23.) which are thought to be indicative for a coarsening in grain-size (Shimmield 1992; Martinez et al. 1999; Moreno et al. 2001; Zabel et al. 2001). While the pre-OAE interval is dominated by cyclic variations, the onset of the black shale is marked by a drop in K/Al, Si/Al and Zr/Al ratios with Si/Al and Zr/Al ratios still remaining constant after the termination of OAE 1b. The synchronous increase in the flux of Al, Si, Zr and K in the OAE 1b interval indicates a significantly higher supply of detrital sediment components to Site 545 compared to pre- and post-OAE 1b conditions (Fig. 25). Sedimentation and accumulation rates will be discussed more in detail in chapter 4.2.5. The dominant process during chemical weathering of the upper crust is the degradation of feldspars and concomitant formation of clay minerals, so that the proportion of alumina to alkalis typically increases in the weathered product. A proxy for the degree of weathering is presented by the chemical index of alteration CIA (1; Fig. 25; Nesbitt & Young 1982, 1996) using molecular proportions:

$$\text{CIA} = [Al_2O_3/(Al_2O_3 + CaO^* + Na_2O + K_2O)] \times 100 \quad (1) \text{ (Nesbitt \& Young 1982)}$$

where CaO* represents only that calcium that is incorporated in silicate minerals (initially used on lutites; Nesbitt & Young 1982) and is not considered in the equation used for this study. Measured primary on volcanic and sedimentary rocks samples with very high CaO contents, indicating the presence of calcium residing in carbonates, were not considered for the calculation of the CIA. The formula calculating the CIA for this study was therefore modified (2).

$$\text{CIA} = [Al/(Al+Na+K)] \times 100 \quad (2) \text{ (data presented in this study)}$$

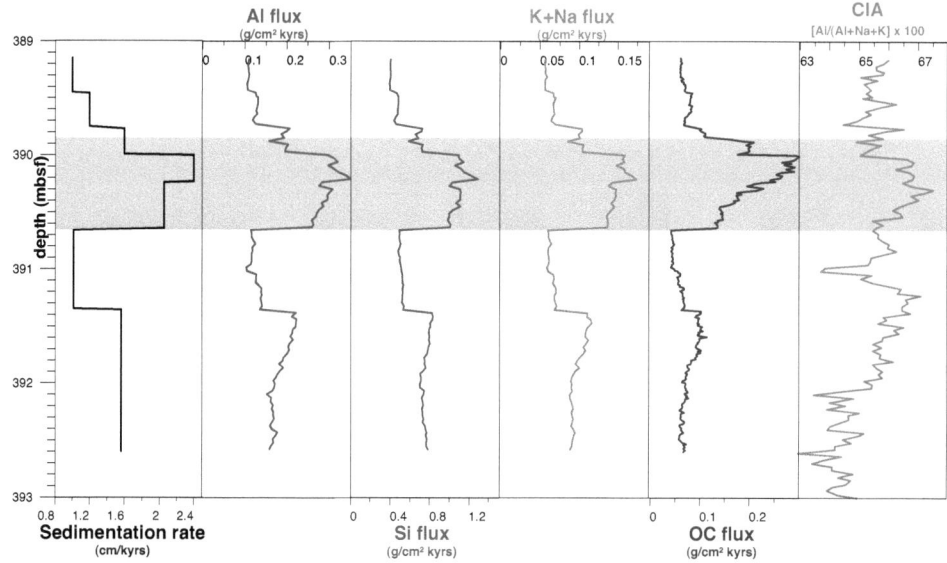

Fig. 25: Sedimentation rates and selected element accumulation rates indicative of the development of the siliciclastic flux and organic matter accumulation at the MP. Sedimentation rates are used for the calculation of accumulation rates. Values resulting from the CIA are discussed in the text.

CIA values from 0 to 50 represent in general no to low degree of weathering, 50 up to 100 accordingly from low to moderate, up to highest degree of weathering. CIA values for Site 545 range between 63-69, lying slightly under the range of average shale values (70-75), suggesting a rather low to moderate degree of weathering for the studied samples. These values result because of the large proportion of clay minerals, like smectite, illite and palygorskite (Nesbitt & Young 1982; Chamley & Debrabant 1984).

Black shales are in general enriched in redox-sensitive trace metals (TM). Enriched sediments are nowadays deposited in coastal upwelling areas and euxinic basins like the Black Sea (Brumsack 2006). In this study the presented redox-sensitive TM are: Ni, V and Zn (as minor elements in ratio to Al $*10^{-4}$). They show only moderate enrichment with respect to average shale values (AVS; Brumsack 2006; Fig. 26) for the investigated interval. V/Al ratios are fairly low at values between 15 - 35. Ni/Al ratios fall slightly above AVS values, with moderate enrichment towards the top of the OAE 1b (Fig. 26). Zn/Al ratios are with few exceptions on average approximately twice as high as AVS in the investigated core section and are not significantly enriched during the OAE 1b interval.

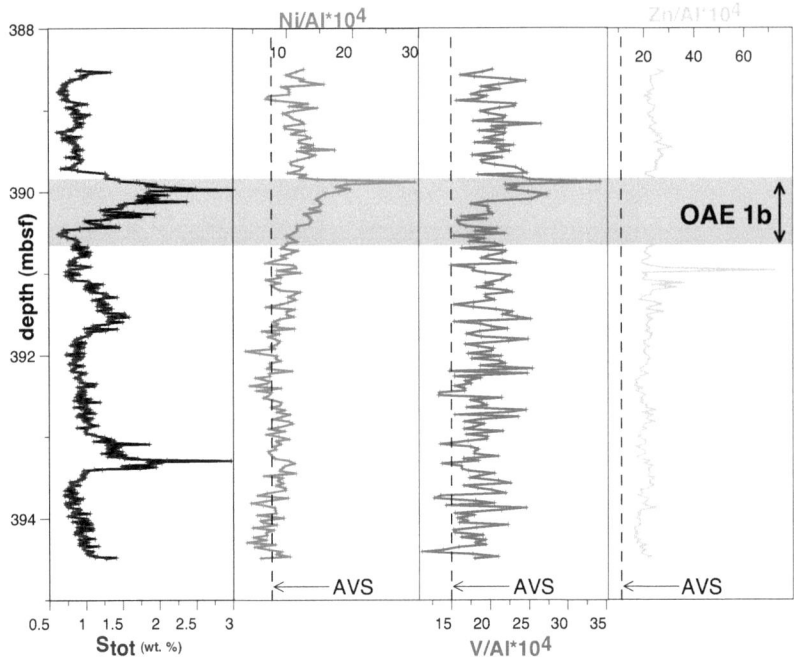

Fig. 26: Variation of selected redox-sensitive elements and element ratios. The broken lines represent average shale values (AVS from Brumsack 2006). Note that the OAE 1b interval is not significantly enriched in redox-sensitive elements.

The aluminum normalised trace element concentrations at the MP are generally low compared to other Albian black shales from the North Atlantic which range e.g., in V/Al ratios from 20 - 200 and Ni/Al ratios from 8 - 80 (Hofmann et al. 1999, 2001), but are similar to trace metal/aluminum ratios from modern upwelling environments (Brumsack 2006). The correlation of Ni and Zn concentrations with TOC content (Fig. 27, a) but not with total sulphur content (S total; $r^2=0.26$ and 0.19 respectively, Fig. 27, b), suggests both elements behaved primarily as micronutrients at Site 545 and were embedded into the sediment via absorption to and complexation processes with organic matter (Tribovillard et al. 2006 and references therein).

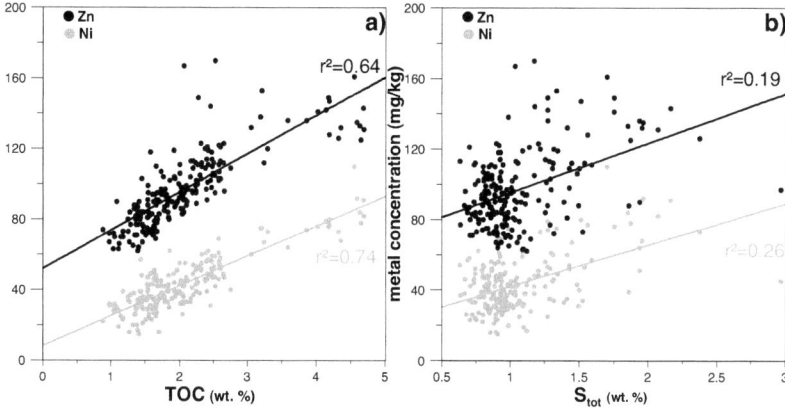

Fig. 27: (a) Correlation of Ni, Zn and TOC at the MP. The good correlation suggest that Ni and Zn acted as micronutrients contrary to the poorly correlation of Ni, Zn and S_{tot} (b).

Anoxic, or even euxinic bottom water conditions which occurred during other Cretaceous OAEs (e.g., OAE 1a and OAE 2; e.g. Kuypers et al. 2004) and resulted in trace metal enriched black shale deposits (Brumsack 2006), were probably not established during OAE 1b at the MP. A drop in the abundance of benthic foraminifera in the OAE 1b interval at the MP (Herrle 2002; Friedrich et al. 2005) indicates somewhat lower oxygen levels of the bottom waters than during the pre- and post-OAE periods even though strictly anoxic conditions with an oxygen content <~0.2 ml O_2/l H_2O were never reached (Algeo and Maynard 2004).

4.2.5. Time model and sedimentation rates

Creating a time model for sedimentation rates was very important for comparing both sites (see therefore also Wagner et al. 2007 and Hofmann et al. 2008). The approach developed by Herrle (2002) was followed, who demonstrated that characteristic shifts in the $\delta^{13}C_{carb}$ curves of the OAE 1b interval can be recognised in all currently best documented OAE 1b sites and, hence, can be utilized for site to site correlation (ODP Site 1049, Blake Nose; DSDP Site 545 Mazagan Plateau and the l'Arboudeysse section of the Vocontian Basin). The Nutrient Index (NI) was used, established for the Vocontian Basin (Herrle 2002) which bases on multivariate statistics and other quantitative analyses of calcareous nannofossils relating taxa to surface water productivity and thus nutrient availability, in order to construct a time model for DSDP 545 at the Mazagan Plateau. The OAE 1b interval of the l'Arboudeysse section displays cyclical variation in nannoplankton productivity (nutrient index) and sediment composition (exemplified by the variation Si/Al ratios) that were linked by Herrle and Herrle et al. (2003) to precession controlled orbital forcing. In the Vocontian study, the positive and negative loadings of different main assemblages given by principal component analyses were used to distinguish and relate cool and nutrient-deplete surface waters from warm and nutrient-rich ones.

The Sedimentation rates (SR) for the OAE 1b interval of the MP are average sedimentation rates for the intervals between correlation points of the carbon isotope curve (Fig. 28). The estimated time content for each isotopically defined interval is based on the precession cycle pattern of the l'Arboudeysse section assuming ~20 kyr for each precession cycle recognized in the nutrient index and Si/Al profiles (Fig. 28).

The LSR (in cm/kyr) was determined following the equation:

$SR\ (cm/kyr) = \Delta h/\Delta t$ were Δt is the difference in depth in cm
and Δt is the difference in time (kyr) between to bench marks

The Vocontian Basin represents the highest sedimentation rates (on average 3.7 cm/kyr) compared to the Mazagan Plateau and Blake Nose and hence the highest time resolution. Bulk sedimentation rates (SR) calculated from the new age model are generally low at the

Mazagan Plateau typical for many Mesozoic open marine black shale settings (Tyson 2005). Bulk SR fluctuate between about 1 cm/ka before and after the event and more than double (~2.4 cm/ka) during the OAE 1b (Fig. 25). The doubling occurred exactly and abrupt at the onset of OAE 1b, values are still increasing during the event and decrease in a staircase shaped pattern with the beginning termination of OAE 1b. At the base of OAE 1b, pre-event values are achieved still declining to obtain lowest sedimentation rates at the end of the record. The entire record presents according to LSR a duration of 250 kyr and OAE 1b formation occurred within ~45 kyr. This duration is in good agreement with previous studies presenting estimations between ~44-46 kyr for the black shale event (see Herrle 2002 and Erbacher et al. 2001).

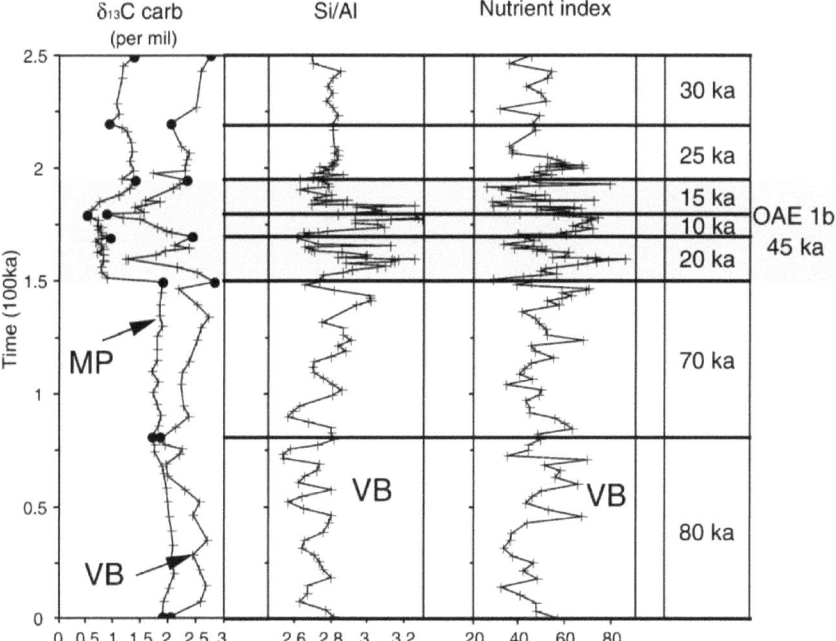

Fig. 28: Correlation and assignment of time content for the Mazagan Plateau and the Vocontian Basin. Black circles on the isotope curves mark the site to site correlation points. The Si/Al and nutrient index profiles from the Vocontian Basin section display a cyclic pattern which is attributed to orbital forcing on the precession scale (Herrle 2002; Herrle et al. 2003). The nutrient index bases on multivariate statistics and other quantitative analyses of calcareous nannofossils relating taxa to surface water productivity and thus nutrient availability. In the Vocontian study, the positive and negative loadings of different main assemblages given by principal component analyses were used to distinguish and relate cool and nutrient-deplete surface waters from warm and nutrient-rich

ones. The time content represented by the sediments between the correlation points is estimated based on the assumption that each cycle recognised in the nutrient index and Si/Al profiles represents ~20 kyr (Wagner et al. 2007).

Organic carbon accumulation rates (OCAR) were calculated according to the following equation:

OCAR $(g/cm^2/yr)$ = sedimentation rate (cm/yr) x sediment density (g/cm^3) x TOC/100

The sediment density (SD) was calculated according to the following equation:

SD (g/cm^3) = [1-(TOC/100)] x 2.7g/cm^3 +(TOC/100) x 1.2g/cm^3

Where 2.7 g/cm^3 is the estimated density of the non-organic matter and 1.2 g/cm^3 the estimated density of the organic matter, respectively (after Stein & Littke 1990; Hofmann 1992).

The presented flux rates of Al, Si, K and OC (Fig. 25), displaying some selected element accumulation rates indicative of the development of the siliciclastic flux and organic matter accumulation, show different trends. At values between 0.05–0.1 $g/cm^2/yr$ the organic carbon (OC) flux remains low before the OAE. At the base of the event this pattern is interrupted by a marked two-fold increase in OC flux that proceeds to almost 0.3 $g/cm^2/yr$ in the centre of the event before it returns to almost pre-OAE levels at the end of the record. With these very moderate rates of OC burial, OAE 1b does not support a major carbon burial event, different from the two main black shale events. The accompanying continental clastic Al and Si fluxes at Mazagan show patterns that closely match variations in SR. Within the first part of the event, however, they are markedly different from the OC flux. OC continuously build up with progression of the event whereas continental fluxes and SR approach almost maximum levels right at the beginning of the event. Although Al and Si are both enriched during the event, the increase of Al is enhanced displayed by the ratio of Si to Al (Fig. 23). This identifies transport modes of the sediment and will be discussed in the following chapters. This temporal relationship indicates high supply of continent matter and nutrients prior to the establishment of conditions supporting enhanced OC burial in the deep ocean. Thus, the Mazagan section does not strictly represent an "Oceanic Anoxic Event" regarding its original definition (Schlanger & Jenkyns 1976).

4.3. Discussion

In the following, the presented and already shortly described results will be discussed. A division into three sub-chapters was chosen, grouping the discussion into a pre-OAE, a OAE and a post-OAE stage. This sub-division should improve pointing to the changes ongoing in a complex climate system during different sedimentation phases.

4.3.1. Conditions prior to OAE 1b

Pre-OAE 1b times are in general characterised by cyclic and moderate warm-cold fluctuations in SST with an amplitude of ~1.5°C (see Fig. 22). One cycle is therefore defined with a beginning warmer and ending cooler state. Following this definition, the record presents two warm-cold cycles prior to the OAE 1b interval. The TEX_{86} derived mean long term temperature range is between ~29 and 31 °C and, hence, similar to SST estimates for the OAE 1b interval at ODP Site 1049 located in the Eastern Atlantic (Schouten et al. 2003; Wagner et al. 2008). High SSTs at the MP reflect the all over significantly warmer global oceanic surface and bottom water temperatures of the mid-Cretaceous oceans (Wilson & Norris 2001; Huber et al. 2002; Schouten et al. 2003; Steuber et al., 2005; Bice et al. 2006).

Cooler intervals are in general characterised by an increase in Si, Zr, and K relative to Al, the warmer intervals generally have high TOC and Al concentrations. Increasing Zr/Al and Si/Al ratios during cooler SST phases at Site 545 may therefore reflect changes in wind intensity and higher transport capacity in combination with the advection of cooler bottom waters during periods of more intense upwelling, which resulted in a rise of the palaeo-thermocline. The overall inverse relationship between SST and wind intensity in the pre-

OAE interval is in good agreement with the existence of a tradewind driven upwelling system off west Africa during the Albian (Leckie 1984; Herrle 2002).

The warmer intervals of the pre-OAE Stage are characterised by the dominant delivery of smectite and kaolinte to the MP, whereas the cooler periods contain higher amounts of illite, heavy minerals and quartz. Studies by Lever and McCave (1983) and work summarised among others in Chamley (1989) demonstrate that the quartz and clay mineral delivery to the Cretaceous Atlantic off North-Western Africa is dominated by aeolian transport. The fluctuation in Si/Al ratios therefore probably reflect changes in the quartz content of the aeolian fraction. The relative enrichment of silica from quartz and elements usually enriched in the heavy mineral fraction (e.g. Zr but also Ti, Y and Cr) compared to aluminum, derived from clay minerals, either suggests that wind strength was highest during cooler climate periods and/or an increase in the formation of clay minerals during warmer interval. A climatic control on clay mineral formation/composition is indicated by the fluctuation in K/Al (see Fig. 23) or illite vs. smectite and kaolinite, and by the variations in Mg/Al ratios that is used as an approximation of the palygorskite/smectite+kaolinite ratio. Illite is generally thought to stem from the breakdown of older rock strata by physical weathering, whereas smectite and kaolinite are generally products of chemical weathering under warm and humid conditions (Chamley 1989). Palygorskite is known from many shallow marine Albian sections along the coastal regions of NW-Africa in association with evaporite minerals such as gypsum and thus probably serves as an indicator for climate conditions with a negative precipitation-evaporation balance (e.g., Pletsch 1998; Pletsch et al. 1996). Marine sediments from the periods of extreme warmth frequently contain palygorskite clay (Pletsch 2003). It could serve as a proxy for warm, saline bottom water when isotope studies are not feasible (Pletsch 2003).

Maximum Mg/Al ratios within the warm/cold cycles of the pre-OAE 1b occur usually at the end of the cooler intervals before they decrease in the warmer intervals (see Fig. 23). These inflection points are interpreted to reflect the dryest conditions, before generally warmer and more humid climate conditions begin to develop as suggested by a relative increase in smectite and kaolinite. High Mg/Al and low K/Mg ratio thus display arid conditions and high Si/Al / Zr/Al ratios indicate cooler intervals. During cooling periods, the aeolian transport, effecting also the coastal upwellling, is accordingly dominant with high transport rates and a charge of increasing grain-size. For warmer intervals more humid conditions, high evaporation rates and at least moderate amounts of precipitation are suggested. During the warmer periods the wind system was weaker compared to the cooler and dryer climate episodes when more quartz, illite and heavy minerals were delivered to the MP (Fig. 29; Herrle 2001; Herrle et al. 2003).

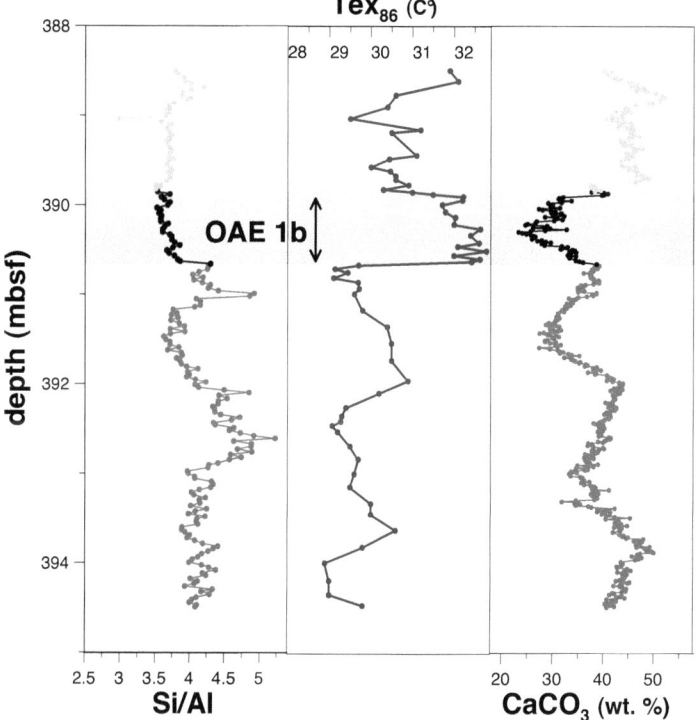

Fig. 29: The comparison of the siliciclastic fraction displayed by the Si/Al ratio with the TEX_{86} record indicates weaker wind systems during warmer periods and a higher delivery of quartz and heavy minerals during cooler episodes when accordingly the wind systems were stronger. Note the different colours marking the different stages for the Si/Al ratio and the carbonate content: dark grey = pre-OAE 1b, black = OAE 1b and light grey = post-OAE 1b.

The TOC content of the sediments of the pre-OAE period is in general low varying from 1-2.5 %. Correlations of organic carbon and Zr content (Fig. 30) further suggest a link of organic matter production, wind strength and upwelling intensity. The organic matter is probably derived from marine and terrestrial sources as indicated by hydrogen indices (HI) around 300 mg HC/g TOC and OI values >125 mg CO_2/g TOC (see Fig. 19) pointing towards poor preservation conditions. A mixture of marine and terrestrial derived organic matter is also supported by the occurrence of marine algael C_{27}-sterenes, sterol ethers and crenarchaeota biomarker at low abundance as well as the presence of long-chain odd-numbered alkanes through the entire investigated interval (Wagner et al. 2008).

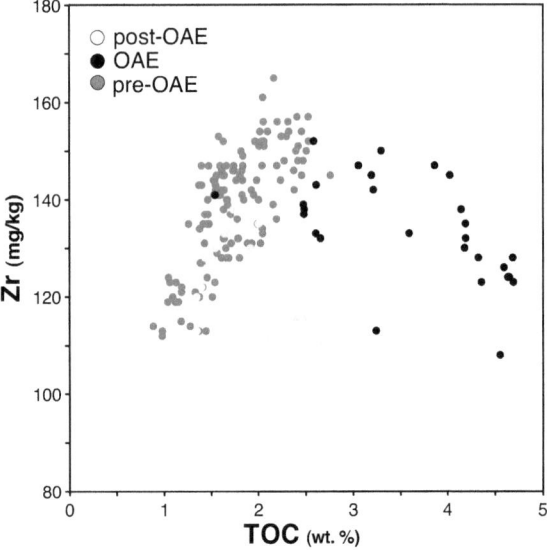

Fig. 30: TOC vs Zr crossplot. Zr is used as a measure for grain-size and wind strength. In the pre-OAE section (dark grey dots) higher TOC values appear with increasing Zr values. This trend is disrupted for the OAE (black dots) and most of the post-OAE interval (light grey dots). Only samples representing the top part of the post-OAE interval reveal the same trend as the pre-OAE sample set.

The good correlation of the S2 pyrolysis data and organic carbon content suggest a fairly constant bulk composition of the organic matter (Fig. 31). TOC content correlates with the detrital siliciclastic fraction, as would be expected for an upwelling environment (Leckie et al. 2002; Handoh et al. 2003). The production of marine organic matter at Site 545 is probably directly linked to upwelling conditions, whereas the terrestrial derived organic matter is more likely to reflect the background supply transported by runoff from the African margin to the adjacent ocean basin. A direct link between the TOC content of the sediments and the warm/cold cycles displayed by the SST data is not evident. Oxygen indices above 100 mg CO_2/gTOC, fairly low Al-normalised concentrations of redox-sensitive trace metals and the presence of benthic foraminifera throughout the OAE 1b

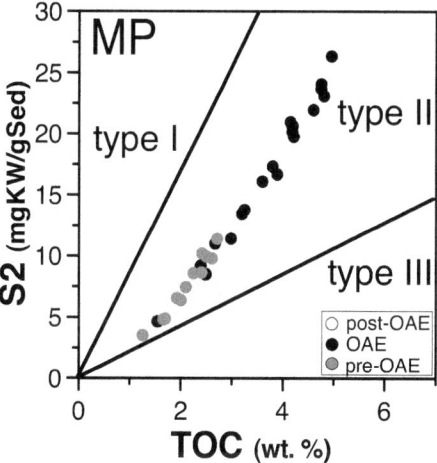

Fig. 31: S2 versus TOC crossplot. Boundaries for Types I, II, and III kerogen fields are taken from Langford and Blanc-Valleron (1990).

containing interval (Friedrich et al. 2005) confirm that the organic matter of the pre-OAE period accumulated in an environment with moderate preservation conditions and at least low amounts of oxygen in the water column. Results from coupled atmosphere-land-ocean modelling (Wagner et al. 2007) support this assumption, suggesting oxygen concentrations in the order of 100 µM. Low sulphur values support this interpretation, even though sulphur fixation was probably limited by the availability of reactive iron as indicated by the good correlation between S_{tot} and Fe content (Fig. 32, A) and in the Fe-S-TOC ternary diagram (Fig. 32, B). It is noteworthy, however, that the maximum Fe and S concentrations are linked to warm climate intervals during the pre-OAE stage, which may be an indication for intensified chemical weathering during these periods on the African continent.

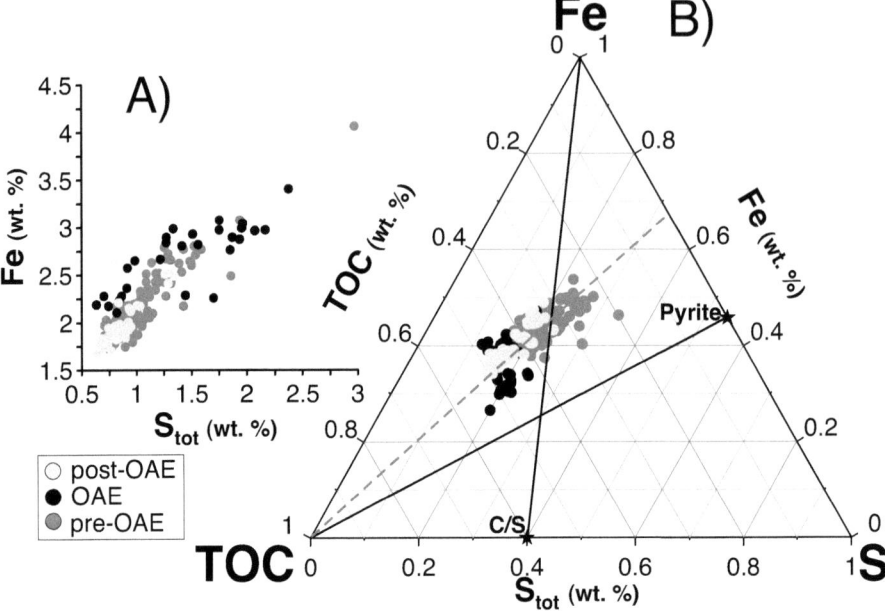

Fig. 32: A) Iron (Fe) versus Sulphur (S_{tot}) crossplot indicating good correlation and pointing to limited S_{tot} due to free reactive Fe. **B)** All the samples in the ternary Fe-S-TOC diagram plot along a line of constant S/Fe with a limited amount of reactive iron supporting the assumed sulphur fixation.

In summary, the pre OAE 1b interval appears to be characterised by upwelling conditions in the North Atlantic sector of the North African continental margin. The upwelling intensity fluctuated in time and seems to be intricately linked to climate fluctuations on the North African continent which are documented by varying wind strength and precipitation intensity.

The concurrent drop in Si/Al, K/Al and Zr/Al ratios in the OAE interval is, hence, most likely caused by a higher supply of fine grained clay minerals at the expense of coarser grained sediment components. The observed accumulation pattern in conjunction with the concurrent shift in sediment chemistry to more aluminium-rich sediment components (clay minerals) suggests a change in the transport processes of the detrital sediment fraction.

4.3.2. OAE 1b at Site 545

The upwelling regime which prevailed throughout the pre-OAE stage was severely disturbed with the onset of OAE 1b. Sea surface temperatures rose rapidly by more than 3.5°C to more than 33°C as indicated by the TEX$_{86}$ data. This increase is more than twice as high as the fluctuation observed prior to the event (in the order of 1.5°C). The observed rise in SSTs by 3.5°C at the MP is an order of magnitude higher than the modelled global temperature rise for OAE 1b of 0.3°C (Wagner et al. 2007), arguing for an amplification of the temperature signal by local feedback mechanisms, such as regionally reduced upwelling. The rise in SST probably occurred within less than 2 kyr if a constant sedimentation rate and a duration of 45 kyr is assumed for the event (see also Erbacher et al. 2001; Herrle et al. 2003).

The observation that the negative shift in the *n*-alkanes slightly preceded that of the marine records (Fig. 22) suggest that ^{13}C-depleted carbon was abruptly released into the atmosphere and subsequently became mixed into surface waters. The algal C$_{27}$-sterene ^{13}C record supports a sudden, though slightly delayed, response in oceanic surface waters relative to the atmosphere. Because the ^{13}C values of *n*-alkanes remained at their depleted levels following the initial pulse, continuous emission of isotopically light carbon for ~25–30 kyr is invoked to maintain this signal for the duration of OAE 1b (Wagner et al. 2007). The divergence between constant $\delta^{13}C_{carb}$ and increasing $\delta^{13}C_{org}$ and C$_{27}$- sterene ^{13}C values probably reflects the evolution of atmospheric pCO_2, consistent with modelling results (Wagner et al. 2007).

The drastic increase in SST was accompanied by the onset of progressive burial of hydrogen-rich OM of marine origin at the sea floor. During OAE 1b the organic matter flux to the sea floor, on average doubled compared to the pre-OAE stage suggesting a substantial increase in surface water productivity and most likely improved preservation conditions (see Fig. 25). A three phase development of organic matter accumulation is recognized for the OAE 1b interval. The initial phase comprises the onset of the negative carbon isotope excursion, a continuous increase in TOC values from 2 to ~3.5%, a gradual increase in HI values from 300 to 400 and a progressive drop in OI values 170 to below 100 (see Fig. 19). The increase in HI and drop in OI values correspond to a gradual reduction in the abundance of benthic foraminifera reported by Friedrich et al. (2005) for this interval and indicates a progressive drop in bottom water oxygenation levels and hence improving preservation conditions for organic matter. The second phase is characterised by TOC values >3.5%, HI values >400 and OI values ~100 as well as a very low abundance of benthic foraminifera (Herrle 2002; Friedrich et al. 2005), suggesting lower bottom water oxygenation levels and good preservation conditions. A moderate increase in S_{tot} values and Ni/Al ratios may also indicate improving preservation conditions.

Comparing SST from the Mazagan Plateau and from North America (ODP Site 1049, Blake Nose) the smaller amplitude in SST change off North America (~2°C) to northwest Africa (~3-4°C) becomes evident and may reflect prevailing upwelling of cooler waters off northwest Africa prior to the event (Wagner et al. 2008). Recalculated SST values from Blake Nose (Wagner et al. 2008; Erbacher et al. 2001) are still as much as 10°C colder (19–24°C) than the TEX$_{86}$-SST data from Site 545 (32–34°C; see Wagner et al. 2008) and substantially lower than modern SST from the area (~27 °C; Levitus et al. 1998) suggesting a strong salinity effect on the $\delta^{18}O$ signal. High surface water temperatures at the Mazagan Plateau probably exclude these strong salinity effects.

The preservation and accumulation of organic matter was probably also enhanced by higher sedimentation rates during phase 2 of the OAE 1b development (Fig. 25). An increase in sedimentation rate effectively decreases the exposure time of organic matter to oxidants (oxygen, sulphate or nitrate) at the sediment water interface and thus reduces the rate of organic matter degradation (Canfield 1989; Hedges et al. 1999). In the third phase representing the termination of OAE 1b, TOC values drop from >4% to less than 2%. Decreasing HI values and progressively increasing OI values, the onset of sediment bioturbation and an increase in the abundance of benthic foraminifera to pre-OAE levels (Friedrich et al. 2005) document the re-establishment of oxygenated bottom water conditions. Even though oxygenation levels fluctuated during OAE 1b at the MP, bottom waters were oxygen deficient during the event, but never anoxic indicated by the relatively low abundance of redox-sensitive trace metals. Modelling results predict a global increase in export productivity during OAE 1b of more than 15 Tmol of marine particulate organic carbon per year accompanied by a moderate reduction in deep ocean water oxygenation by about 10 µM to concentrations above 90 µM (Wagner et al. 2007) suggesting that both enhanced organic matter production and improved preservation conditions fostered the enhanced accumulation of organic matter during OAE 1b.

The increase in SST and marine organic matter accumulation during OAE 1b at DSDP Site 545 was probably not caused by intensification of local upwelling conditions, even though the flux of alumosilicates almost doubled (Al, Si and K flux; Fig. 34). Si/Al, Zr/Al and K/Al ratios and, hence, wind intensity is expected to have markedly dropped for the North African region during the event (Fig. 33).

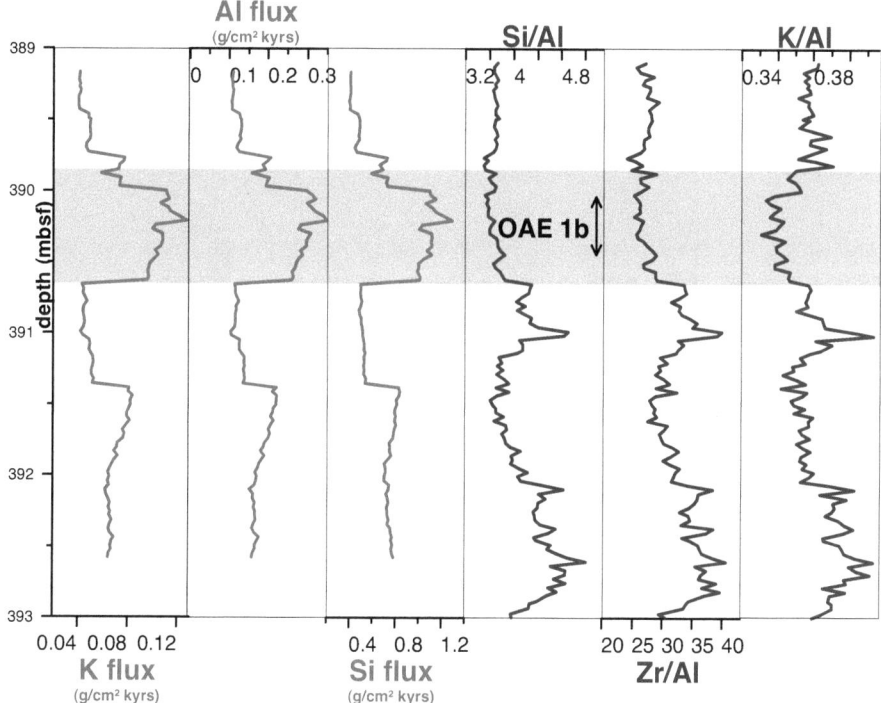

Fig. 33: Flux rates for potassium (K), aluminium (Al) and silicium (Si) show an abrupt rise at the onset of OAE 1b sedimentation whereas the ratios normalised on aluminium are slowing down, pointing to higher Al concentrations during the event.

The co-variation between organic matter and Zr content, characteristic for the pre-OAE interval, is not observed during the OAE 1b interval (see Fig. 30) supporting a decoupling of organic matter production from wind strength and resulting upwelling intensity. Preferred delivery of Al-rich clay minerals (low Si/Al, Zr/Al and K/Al ratios) in conjunction with rising

SSTs during OAE 1b suggest warmer regional climate conditions leading to an accelerated hydrological cycle which fostered precipitation along the African continental margin and enhanced neo-formation of clay minerals. The attenuation or break down in (trade) wind strength for the North African region may have triggered a northward expansion of monsoonal circulation into subtropical latitudes stimulating even higher amounts of continental run off and enhanced the delivery of nutrients and clastic sediment to the adjacent ocean basins (Herrle et al. 2003). At the MP elevated nutrient fluxes resulted in a doubling of the sedimentary phosphorous flux during OAE 1b (Wagner et al. 2007). High amounts of precipitation and fluvial discharge during the event are also reported for the North American continental margin (Erbacher et al. 2001) and the South France Basin (Herrle 2002; Wagner et al. 2007), in line with the presented results.

A shift of the ITCZ (Inter Tropical Convergence Zone) towards the MP could have fostered a shift in the air-sea freshwater flux (equivalently precipitation minus evaporation mass balance, P-E) leading to higher rainfall and a positive P-E mass balance. Site 545 was located at a palaeolatitude nearby 20°N presenting nowadays as well as during Albian times a negative P-E mass balance (Ufnar et al. 2004) which was apparently disrupted during the formation of OAE 1b.

4.3.3. The post OAE stage

The termination of OAE 1b is marked by an almost synchronous drop in SSTs (approximately 2°C) and TOC concentration, and a 0.5‰ shift towards heavier carbon isotopes in bulk carbonate. However, sea surface temperatures remained warmer than during the pre-OAE period and the perturbation of the upwelling system prolonged the OAE 1b event (see Fig. 29). Detrital flux levels similar to pre-OAE values are re-established approximately 25 kyr after the termination of the black shale. The generally weak trade wind system, characteristic for the OAE stage, persisted in the post-OAE stage as indicated by continuously low Si/Al, Zr/Al and K/Al ratios (Fig. 34). Pre-OAE levels of Si/Al and Zr/Al are only reached in the top 30 cm of the investigated interval where increasing TEX$_{86}$ SSTs suggest the onset of a new temperature cycle which is, however, only partially recorded in core 42. The top 30 cm of the core display TOC/[Zr/Al] relationships similar to that of the pre-OAE stage were increasing TOC values correspond to increasing Zr/Al ratios (see Fig. 30) suggesting an overall return to pre-OAE upwelling conditions for this interval.

The observations described above, show that the perturbation of the regional climate system over NW Africa which resulted in the attenuation of the trade wind system and weakening of upwelling conditions, thus persisted considerably longer than the isotope excursion associated with OAE 1b, providing evidence for a significant lag time in the response of the Cretaceous land–ocean climate system to atmospheric heating.

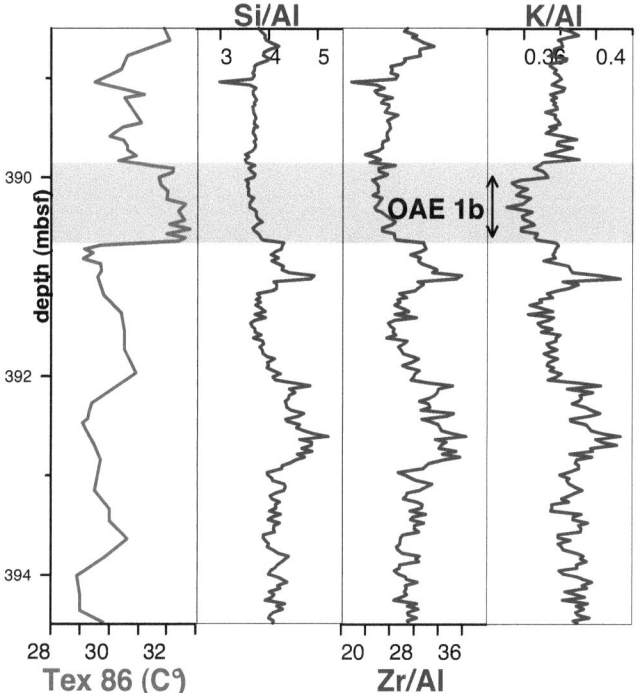

Fig. 34: The break-down of the upwelling system introduced by the OAE 1b sedimentation continued in the post-OAE stage when temperature began to reduce again. Returning to normal upwelling conditions takes almost the same time as the duration of the black shale.

A return to pre-OAE upwelling conditions may have occurred approximately 25 kyrs after the event, but the reasons and mechanisms for this apparent time lag are currently only poorly understood. It can only be suggested, that the still weakened trade wind system may have resulted in continuously moist and warm conditions over the African continent driving intense weathering and neoformation of clay minerals and an ongoing flux of aluminium-rich sediment to the North Atlantic. The weakening of the trade wind system may still be a result of a positive precipitation balance caused by a shifted ITCZ.

After presenting and discussing in the following chapter the data from the Vocontian Basin, predictions about climatic scenarios affecting palaeoenvironments at regional and global scale and mechanisms causing OAE 1b will be discussed and summarised in the Synthesis (Chapter 6).

5. Vocontian Basin

5.1. Palaeogeography and Geology

In France and surrounding areas, outcrops of the outer alpine area comprise: (1) basement uplifts, where pre-Triassic rocks crop out and (2) Mesozoic-Cenozoic sedimentary basins with nearly non folded Mesozoic and Cenozioc sediments (Arnaud & Lemoine 1993). In these areas the Alpine belts are the Pyrénees and their Eastern extension in the Provence ranges and the Western Alps. Three main sedimentary basins can be distinguished: (1) The Paris Basin, (2) the Aquitane basin and the South-East France Basin (SFB). The SFB is a Mesozoic Basin that is located between the internal zones of the Western Alps and the outer-Alpine French Central Massif. The western part of the SFB does not belong to the folded Alps. Its eastern part belongs to the External zone of the Western Alps and its southern part belongs to the Alpine folded Pyrenean-Provence belt (Fig. 35).

Sediments presented in this study belong to lowermost structural units from the External zone called Dauphíne or Dauphinois in the Western Alps. They originate from the European margin of the Mesozoic Ligurian Tethys ocean. The Western Alps are divided in two major parts, the Internal and the External Zone. The Internal Zone represents the Penninic nappes and the Austro-Alpine. The External Zone comprises a parautochthonous (Ultra-Dauphiné, Ultra-Helvetic) and an autochthonous part (Dauphiné, Helvetic; Lorenz 1980). The whole External Zone was part of the European margin of the Mesozoic Tethys ocean (Arnaud & Lemoine 1993). Because of the more marginal position of the Dauphiné at the realm of the Alpine fold belt, this unit is less deformed than the Ultra-Dauphiné (Gwinner 1971). A second major structure in SE-France is the South-East France Basin (SFB). The study area in SE-France, the Vocontian Basin or French subalpine basin, belongs to the Dauphiné part of the SFB.

Some authors (Lemoine 1984, 1985) consider the French subalpine Basin as an analog of the present European margin of the Atlantic Ocean, for it was formed by a series of tilted blocks that deepened eastward toward the open Tethyan Ocean. Ferry (1990) suggest that the evolution of the area was in relation with an aborted rift basin (Aulacogen). In this view, the French subalpine Basin may be considered as a sort of pull-apart basin. Three subsidence phases are recognized within the Early Cretaceous of the Dauphinois Basin: (1) a Valanginian–Hauterivian phase characterised by a relatively uniform subsidence pattern; (2) a Barremian–Gargasian phase characterised by profound shifts in subsidence; (3) a Clansayesian–Albian phase characterised by relatively small rates of tectonic subsidence or uplift (Fig. 36). The Barremian – Gargasian tectonic phase, with the Early Barremian uplift of the entire Dauphinois Basin finally resulting in Gargasian exposure of the margins and accelerated subsidence of the basin centre, is associated with the changing tectonic regime in the Piemonte–Ligurian ocean from extension towards compression, thus forcing the onset, progradation and disappearance of extensive carbonate platforms at the margin of the basin (Wilpshaar et al. 1997).

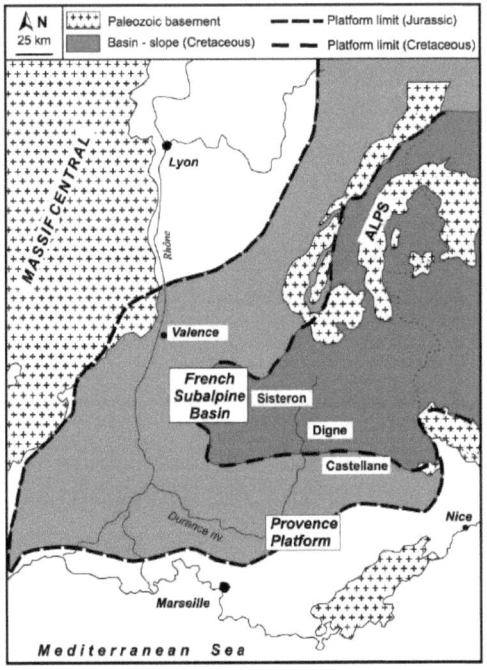

Fig. 35: The Vocontian Basin was bounded by the crystalline basement of the Massif Central to the West, by the Provençal Platform to the South and it was open to the East towards the Tethys Ocean. Platform boundaries for Jurassic and Cretaceous are shown (Olivero D. & Mattioli E. 2008).

The Vocontian Basin was bounded by the crystalline basement of the Massif Central to the West, by the Provençal Platform to the South (represented now by the Corsica-Sardinia and Maures-Esterel massifs) and it was open to the East towards the Tethys Ocean. In the Middle Jurassic, the central part of the basin was subject to strong subsidence, evinced by marls formed under dysoxic conditions ("Terres Noires" Formation). A marked regression occurred during the Late Jurassic (Atrops 1984), when calcareous sediments indicative of a shelf environment were deposited within the entire basin. From Berriasian through Aptian times, in a general regime of extensional tectonics, the depth of the basin increased progressively. Limestone-marl alternations were deposited (Atrops 1984). A maximum depth was probably attained during the Aptian-Albian interval. At the end of the Albian, sediment accumulation in the basin started, main phase of emergence took place during the Santonian. This uplift is contemporaneous with the first phase of Alpine tectonics: the "Pyrénéo-Provençale" phase. This phase involved the formation of structures with a general W-to-E trend.

During the Jurassic the SFB was surrounded by numerous carbonate platforms displaying a triangular shape which has been reduced in the Early Cretaceous as a result of the shift of its North-Western and South-Western platform limits (now the Urgonian platform). This is the time of the Vocontian Period (Latest Hauterivian-Cenomanian) which includes the deposition of the Cretaceous black shales until the final closure of the basin during the Tertiary. During the Late Jurassic and Early Cretaceous the epicontinental Vocontian Basin was a part of the SFB and was located at a palaeolatitude of 25° to 30°N (Savostin et al. 1986; Voigt 1996; Hay et al. 1999). During the Vocontian Period the deep basinal part was relatively narrow, being surrounded by the Urgonian limestone platforms. The shallow-water carbonates of the surrounding and later drowned platforms and slopes intercalated with the hemipelagic facies in the centre of the basin. To the east, the Vocontian basin was open to the Tethyan Ocean (Curnelle & Dubois 1986; Arnaud & Lemoine 1993). From Aptian-Albian two major events occurred: (1) the submersion of the Urgonian platforms and the end of the carbonate sedimentation, and (2) the birth of a siliciclastic sedimentation (Arnaud & Lemoine 1993). This change resulted not only from the rising sea level but had also climatic causes.

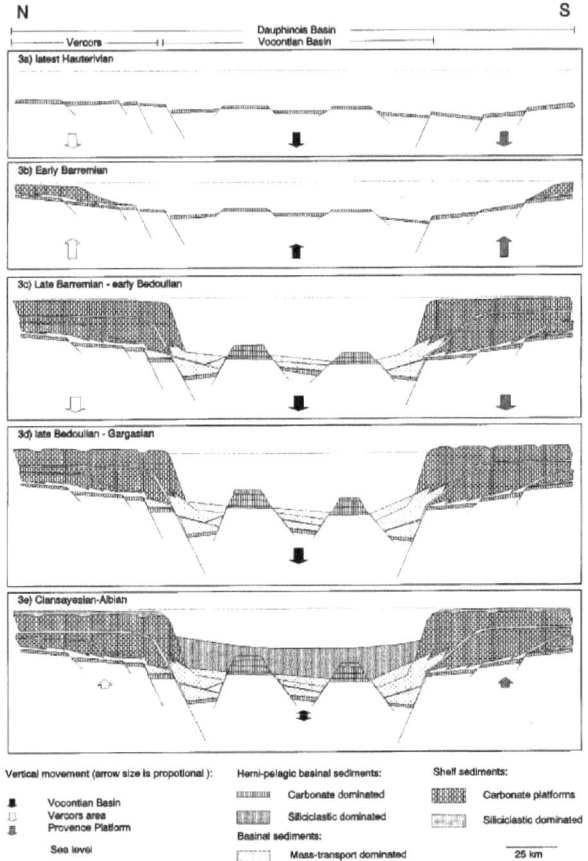

Fig. 36: Simplified diagram (Wilpshaar et al. 1997 and references therein) illustrating the tectonic scenario for the latest Hauterivian-Albian interval. Vertical scale exaggerated.

During these transgressive intervals the deposition of very thick lime- and marlstones (up to 800 metres) proceeded into the Vocontian through. In the western part, including the study area, the series contain re-sedimented glauconitic sands and several organic-matter rich shales, representing several superimposed depositional sequences. Gravity reworked green sands reflect lowstand system tracts while maximum floodings are decimetric to metric black shales levels and highstand system tracts are marly, often less thick than the lowstand system tracts still containing very thin black shale levels.

Palaeo-water depths for the Vocontian though are a fact of discussions. Based on faunal and floral data, the palaeo-water depth of the Vocontian Basin is estimated by some authors within a range of several hundred metres (Wilpshaar & Leereveld 1994; Wilpshaar et al. 1997). In contrast, other authors recommend a palaeo-water depth up to 2000 m for the Vocontian Basin (Cotillon & Rio 1984), reflecting middle to lower bathyal water depths (1000-2000 m) during the late Albian (Holbourn et al. 2001). The study from Holbourn et al. (2001) based on benthic foraminiferal assemblages support the existence of a relatively wide and deep-water connection between the eastern and the western Tethys (>2500 m) through the Gibraltar Gateway during the late Albian. Other (Ferry 1990) suggest that the Vocontian Basin probably attained its maximum depth (~500 - 800 m) in the Early Cretaceous (Hauterivian-Barremian).

In summary, the SFB including the Vocontian Basin represented in general a deep pelagic basin (Arnaud & Lemoine 1993) surrounded by drowned carbonate platform areas to the West and South and the Tethyan Ocean to the East. It represents due to its position high tectonic activities.

5.1. Results

The achievement of the analysed sediment record are in the following presented in distinct sections in the same order as earlier presented for the Mazagan Plateau. In the respective sub-chapter the results will be described in detail and shortly discussed. The main data set will be presented in either xy- or triangle-plots against depth. Due to the technical approach, presentation of the organic matter will include gas chromatograms and HI/OI diagrams. The ensuing discussion chapter (see 5.3.) highlights a summing up picture of the data record, putting together all results and presenting them in a threefold system before-, during- and after the OAE 1b event.

5.2.1. Stable carbon isotopes

The OAE 1b interval can be defined by the bulk carbonate and bulk organic carbon $\delta^{13}C$ isotope records (Fig. 37). The entire record presents fluctuations which are approximately enhanced during OAE 1b showing stronger contrasts. The black shale interval starts with a negative isotope excursion. Strikingly the isotope shift is two-folded, consisting of two negative pulses or spikes. The OAE 1b interval can therefore not be seen as a constant negative isotope excursion as the one at the MP although the general magnitude remains negative. After the first negative impulse, values are becoming lighter in the central part of the black shale before they return, due to the second spike, towards heavier values and then decrease again slightly towards the end of the event.

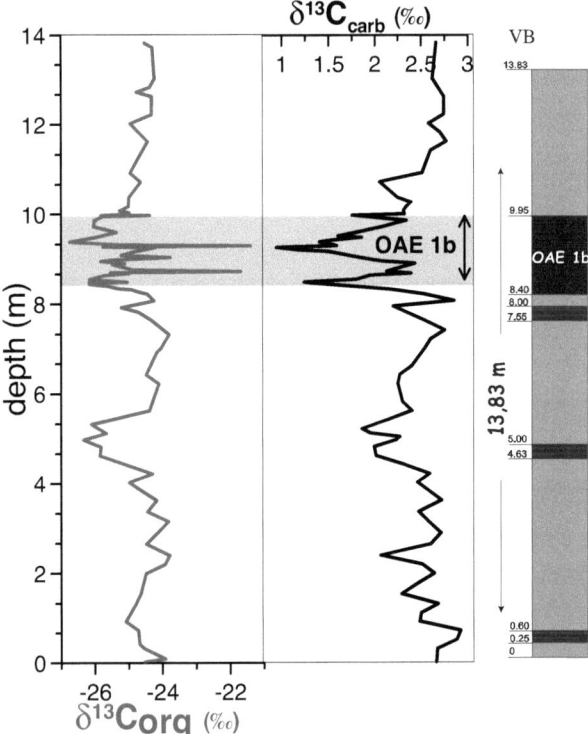

Fig. 37: Bulk stable carbon isotopes from the Vocontian Basin. Bulk carbonate $\delta^{13}C$- isotopes extended after Herrle 2002. The grey shaded interval marks the duration of OAE 1b. Note, the negative shift reveals two peaks and displays various fluctuations. The profile on the right hand side is taken from Fig. 12.

Pre-OAE- $\delta^{13}C_{carb}$ values fluctuate between ~2 ‰ and ~3 ‰ showing a firm swinging pattern. As a reminder, the time model presented for the MP is based on the precession cycle pattern of the l'Arboudeysse section (see Chapter 4.2.5. for details). The $\delta^{13}C_{carb}$ record show a slightly swinging pattern representing 100 kyr cycles, visible in the bulk carbon isotopes (Fig. 37). Values become slightly more positive at the base of OAE 1b (around 3 ‰) before they shift into a negative pulse of ~1.2 ‰ at the beginning of the black shale (at meter 8.40). The shift proceed progressively and not as abrupt as at the MP, occurring within 40 cm (and before the onset of the black shale sedimentation) presenting thus almost ten thousand years. In the same magnitude, values increase and decrease again. This pattern is very steady and excludes, at first sight, extraordinary one-time events. Here, the OAE 1b is more comparable to most of the Cretaceous black shales which show a short pulse negative shift followed by a positive excursion (e.g. OAE 1a and OAE 2). But the positive excursion is missing because values are becoming again even more negative (~1 ‰) in the middle part of the event and then drop slightly in the top third of the event to pre-OAE values (~2.2 ‰). The record commences after the event in the same magnitude as prior to it. The termination phase of OAE 1b is comparable to Site 545, a roughly steep but gradual return to isotopic values of 2.3 ‰ by a drop of 1.4 ‰. A short disturbance or recovery phase to more negative values is evident after the event, but than the cyclic fluctuation pattern is quickly re-established.

The ^{13}C record of OC ($\delta^{13}C_{org}$) parallels that of $\delta^{13}C_{carb}$ with only minor differences. During the pre-OAE in the lower part the variable swings are more obvious, presenting a cyclic pattern fluctuating from under –24 ‰ up to –26 ‰. The post-OAE is therefore less pronounced in the swinging magnitude as the $\delta^{13}C_{carb}$ values remaining constant around –24 ‰ recording same general pattern as prior to the event. The onset of the black shale is marked by a negative shift of only ~0.3 ‰. Same as discussed before for the carbonate carbon isotope record, the shift from the most positive value of –24.2 ‰ to the negative peak of –25.6 ‰ at the onset of OAE 1b occurred around 40 cm prior to the event. Within the event, both isotope records follow roughly the same pattern and remain clearly within same magnitude. Two unfitting $\delta^{13}C_{org}$ -values are extremely positive around –21‰ without a match in the corresponding $\delta^{13}C_{carb}$ record. An obvious explanation is missing as there is also no recurrence in the entire geochemical record. Apart from that, $\delta^{13}C_{org}$ values within the black shale are in general around –25 ‰, with mean negative values of –26 ‰ and mean positive values around –24 ‰ and decrease likely to the $\delta^{13}C_{carb}$ record in the upper third of the event termination.

There is no obvious evidence for constant supply of ^{13}C-depleted CO_2 during the entire black shale interval at the Vocontian Basin because the negative shift initiate earlier to OAE 1b and decreases gradual before a second negative pulse commence. Values than decrease again gradual towards the termination of the black shale. The fluctuations of both isotope records show that general trends are somehow enhanced during the event presenting stronger contrasts.

5.2.2 Organic composition

TOC values before and after OAE 1b at the Vocontian Basin fluctuate between 0.7 to 2.8 % (Fig. 38). Highest values, excluding the OAE interval, are reported in the two black shale layers pre-dating the OAE 1b (see 2.1. for details). Suspending these layers, TOC value fluctuations are in the range of nearly 1 % (0.7%-1.8%). In contrast to Site 545 there is no superordinate cyclicity visible. Here, small scale variations are dominant. Post black shale values are therefore almost stagnant around 1.2 %. Very similar to Site 545 the organic matter content increases directly prior to OAE 1b by 0.6 % with increasing tendency achieving a rise of almost 2 % (TOC around 3.4% at meter 8.49) within ten centimeters. The OAE 1b interval is characterised by a shift to higher TOC values remaining constant during the entire interval with an average mean of 2.8 % (varying between 2%-3.8%).

The shift in TOC content is evident but not as pronounced as at Site 545 and values up to almost 5% are missing. During the black shale, a constantly elevated TOC reveals small scale variations, very similar to pre-OAE 1b conditions. A rapid return to TOC contents similar to pre-OAE values, occur in the first ~5 cm after the event. From all analysed parameters the TOC record traces the clearest contour of OAE 1b with its sharp beginning and termination supporting the carbon isotopes. This might help to classify other parameters in the following.

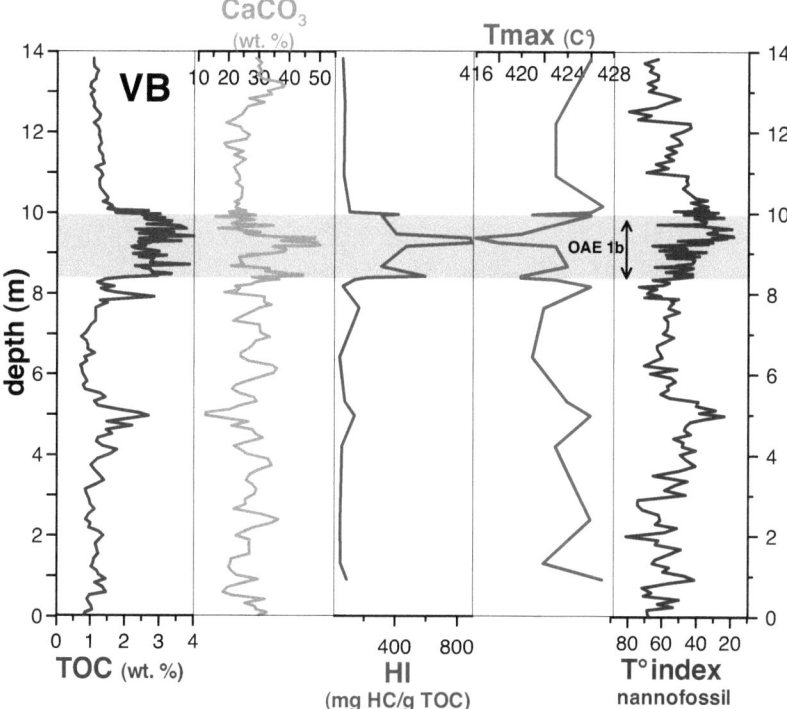

Fig. 38: Geochemical record of the complete analysed section from the Vocontian Basin. Organic composition: TOC, HI and Tmax. Furthermore presented is the carbonate carbon content (CaCO₃) and the temperature index (TI, after Herrle 2002) based on calcareous nannofossil assemblage.

Hydrogen indices (HI) follow changes in TOC and range from 37 to very high 886 mg HC/g TOC values (Fig. 38). HI values around 100 HI mg HC/g TOC are measured for the entire record, except for the OAE 1b interval and the other three black shale intervals. HI values show a sharp increase to highest values at the beginning of the OAE 1b interval with a sudden decrease at the top of OAE 1b and reveals a similar pattern to TOC content. Plotting the data in a Hydrogen index *versus* Tmax diagram (Fig. 39; after Langford & Blanc-Valleron 1990) the values indicate that the organic material has a broad mixed kerogen character of type II/III (a mixture of marine and terrestrial material). At the second look a clear classification of the material is simple, as samples of kerogen type II belong only to the OAE 1b interval and the ones of type III belong to the remaining record (Hydrogen indices below 100 mg HC/g TOC with values up to 850 mg HC/g TOC within the black shale). The organic material of the Niveau Paquier therefore is of predominantly marine origin. Pyrolysis temperatures (Tmax, Fig. 38) below 425 °C indicate that the OC is immature although the composition of the black shale organic matter indicates thermal degradation, i.e. it is thermal mature (Vink 1996).

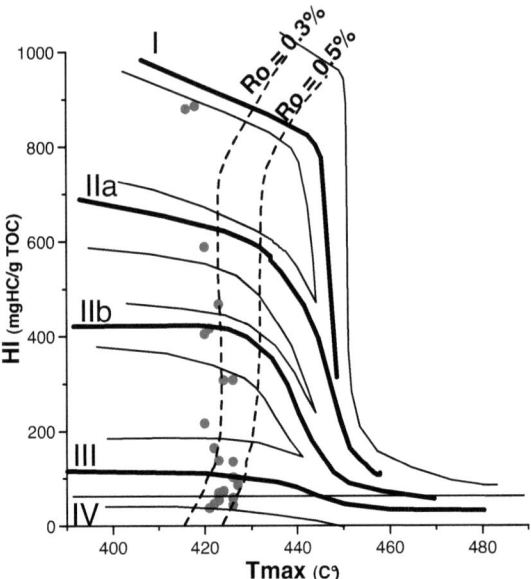

Fig. 39: Samples from the Vocontian Basin plottet in an hydrogen index *versus* Tmax diagram (after Langford & Blanc-Valleron 1990).

At the Vocontian Basin the composition of the OM is of isotopically heavy archaeal origin, characteristic for non upwelling OAE 1b sites in the western Tethys and at Blake Nose (Kuypers et al. 2001; Tsikos et al. 2004). The SFB (South-East France Basin) represented a deep pelagic basin (Arnaud & Lemoine 1993) surrounded by drowned carbonate platform areas to the West and the Tethyan Ocean to the East. It was to some part a restricted basin but with a diverse source of OM delivery dependent certainly also on deep water formation. Here, up to 40 wt % is of marine, chemoautotrophic archaeal OC contribution (Kuypers et al. 2002a; Tsikos et al. 2004). In general the source of OM is a mixture of largely algael-, -bacterial and archaeal derived biomarkers (see 5.2.3.). The abundance of OM derived from archaea during OAE 1b is unexpected and matchless marking a unique period in Earth history and indicating totally different mechanisms for OM deposition than has been invoked for other OAEs (Kuypers et al. 2001).

Different to Site 545 the carbonate carbon record ($CaCO_3$) for the VB is during the black shale event not inversely to TOC (Fig. 38). Concentrations vary between 12 and 50 % $CaCO_3$ with highest values during the OAE 1b and an average mean of 28 %. The pre-OAE interval is characterised by cyclic carbonate fluctuations on a small scale variation, representing precession cycles. With the onset of OAE 1b the $CaCO_3$ content abruptly increases up to 40 % but decreases in succession again, prevailing the same magnitude as prior to the event. The next highstand of $CaCO_3$ content yielding up to 50 % is furthermore the last one during OAE 1b which concludes with decreasing but still strongly contrasting values (Fig. 38). Post-OAE values remain constant at the base with values around 23 % resuming small scale variations in the upper part of the record.

The total abundance of calcareous nannofossils reported by Herrle (2002) correlates with the TOC content. Low total abundance of calcareous nannofossils within the OAE 1b corresponds to high abundances of *Nannoconus* spp. (see Herrle 2002 for more information). Highest simple diversities occur within the Niveau Paquier black shale and shortly above the two other black shale horizons (at 0.75 m and 5.05 m). During the formation of OAE 1b three short repopulation events of benthic foraminifera can be recognised, reflecting high-frequency temperature changes (Friedrich et al. 2005). The carbonate trend correlates with the nutrient index (NI; described by Herrle 2002) reflecting the same fluctuations and intensities. In general the carbonate record displays precessional cycles with constant variations. During the OAE, these fluctuations remain stable displaying enhanced magnitudes. Different to the TOC content, low carbonate values comparable to the prior-OAE content, are present. The rhythmic pattern is interrupted at the base of the post-OAE data set but is re-established at the top of the record.

A very diverse and varying sulphur content reflects the cyclic variability of the presented record. The total sulphur (S_{tot}) content of the entire section ranges in the order of 0.5 % (Fig. 40). Highest contrasts in sulphur concentration are documented during the OAE black shale but start earlier (at metre 8). The trend is only apparently comparable to the carbonate content as the sulphur content is usually high when carbonate is low. But a short reversal is obvious during the Niveau Paquier when highest sulphur values are reached (up to 0.75 %). As documented for Site 545, higher sulphur concentrations are here in general recorded during cooling episodes following warmer episodes as indicated by the temperature index (TI; Fig. 38, Herrle 2002). Plotting sulphur and iron (Fe) in an xy-plot, the data set is strongly scattered without preferential direction (Fig. 41).

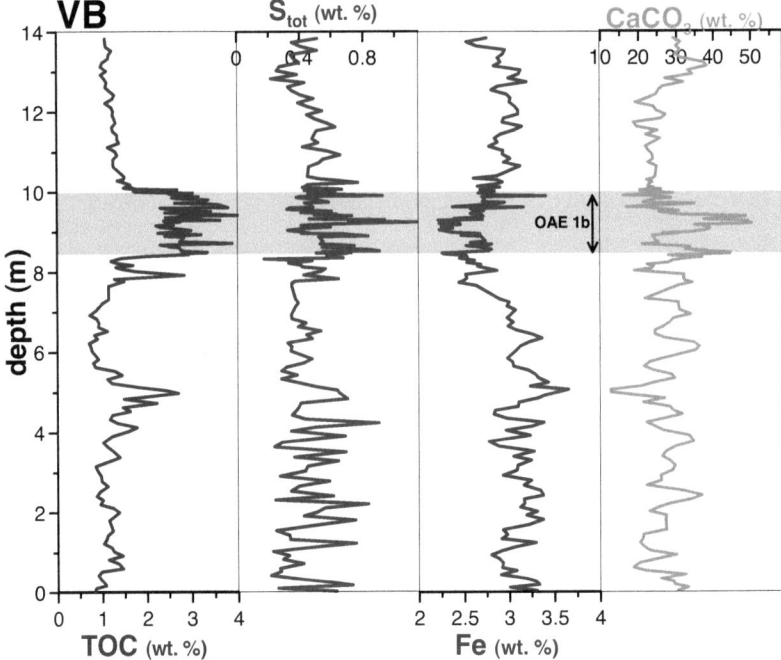

Fig. 40: Geochemical record of the total organic carbon- (TOC), the total sulphur- (S_{tot}), the iron- (Fe) and carbonate (CaCO3) content.

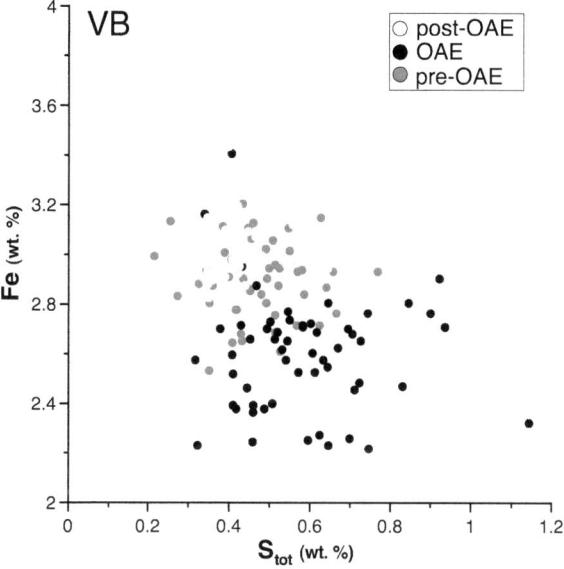

Fig. 41: Correlation of Iron and S_{tot}, showing a scattered field without preferred direction.

Summing up, highest Fe- and lowest sulphur content are documented after the OAE, lowest Fe- and highest sulphur content are recorded during OAE and the record prior to OAE presents a broad mixture of both. The ternary TOC-Fe-S_{tot} diagram (Fig. 42) does not suggest iron limitation for the fixation of sulphur in the sediments (Raiswell & Berner 1985) as described for Site 545 record. Here most of the samples plot in the ternary TOC-Fe-S_{tot} diagram along a line extending from the TOC corner to the Fe-Stot line indicating that sulphur and iron availability, strongly depended on TOC concentrations. The samples have in general lower sulphur, slightly higher iron- and TOC content than the Mazagan record.

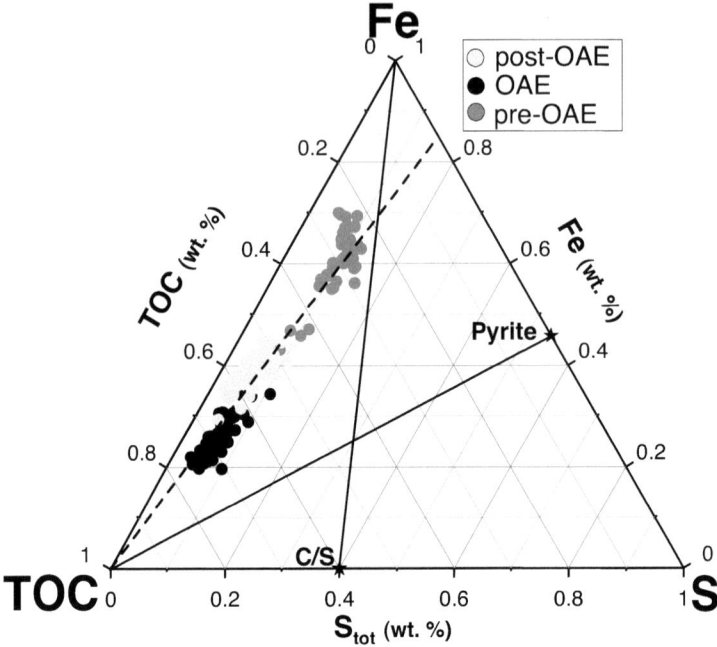

Fig. 42: Samples from the Vocontian Basin plot in the ternary Fe-S-TOC diagram along a line extending from the TOC peak to the Fe-S_{tot} line. Contrary to Site 545, here iron availability depended on TOC concentrations.

5.2.3. Biomarker

The composition of the OM is mainly marine origin. The apolar fraction of the entire record contain short-chain *n*-alkanes (C_{21}-C_{25}) with no odd-over-even carbon number preference (OEP1 ~ 0.9). Long-chain (C_{25}-C_{29}) *n*-alkanes show a low odd-over-even predominance (OEP2 ~ 1.7) suggesting that leaf waxes of terrestrial plants are a possible source for the OM. CPI (CPI = 0.2-3.5) values show stronger variations which are in good correlation with changes in TOC (Fig. 43). CPI or OEP values below 1.0 are in general unusual and typify low-maturity oils or bitumen from carbonate or hypersaline environments. CPI values below 1.0 are present in the pre- and post-OAE record, during the event values range from 1.5-3.8. Long-chained n-alkanes derived from terrestrial plant leaf waxes (C_{27}, C_{29}, C_{31} n-alkanes) are present outside of the OAE interval and were also determined by compound-specific carbon isotopes measurements. Unfortunately non of these n-alkanes exist in measurable concentration within the OAE. Main composition of the OM is derived from prokaryotic (i.e. bacterial/cyanobacterial) and eukaryotic organisms. The components include mixtures of acyclic isoprenoids, n-alkanes, hopanoids and steroids (see also Vink et al. 1998). The high relative abundance of acyclic isoprenoids in the hydrocarbon fractions during the black shale sedimentation suggest that methanogenic archaebacteria were present at the time of deposition.

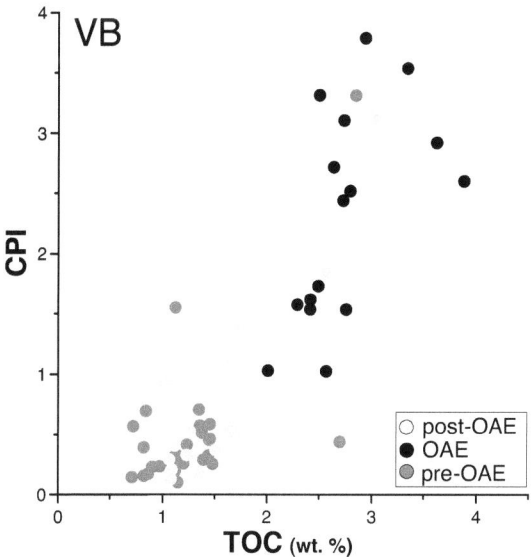

Fig. 43: CPI (Carbon Preference Index after Bray & Adams 1961) versus TOC crossplot.

In recent and ancient sediments the irregular C_{25} isoprenoid skeleton (2,6,10,15,19-- pentamethylicosane; PMI) has been used as a marker of methanogenic activity as it has only been found in methanogenic archaea (Vink et al. 1998). Two saturated isoprenoids compounds containing 24 and 26 carbon atoms and apparently closely related to PMI (Vink et al. 1998; Kuypers et al. 2002a) are also present in the OAE interval, namely the

C_{24} isoprenoid (2,6,15,19-tetramethylicosane;TMI) and the C_{26} isoprenoid (10-ethyl-2,6,15,19-tetramethylicosane;ETMI). Both are suggested to indicate an common archaeal origin (Vink et al. 1998).

On a total ion current (TIC) these isoprenoids elute between the C_{22} and C_{24} n-alkanes (Fig. 44). They are also reported from the Ionian Basin, NW Greece (Tsikos et al. 2004) and from ODP Site 1049 in the North Atlantic (Kuypers et al. 2002a). At the record of the VB these isoprenoids prevail in high abundance but are restricted to the OAE interval and its nearby bordering sediments. PMI and TMI concentration range in equal parts, with apparently no significant isoprenoid preference. The sedimentary compound ETMI is present in lower concentrations and the identification is difficult because of co-elution with the regular C_{23} n-alkane (Fig. 45).

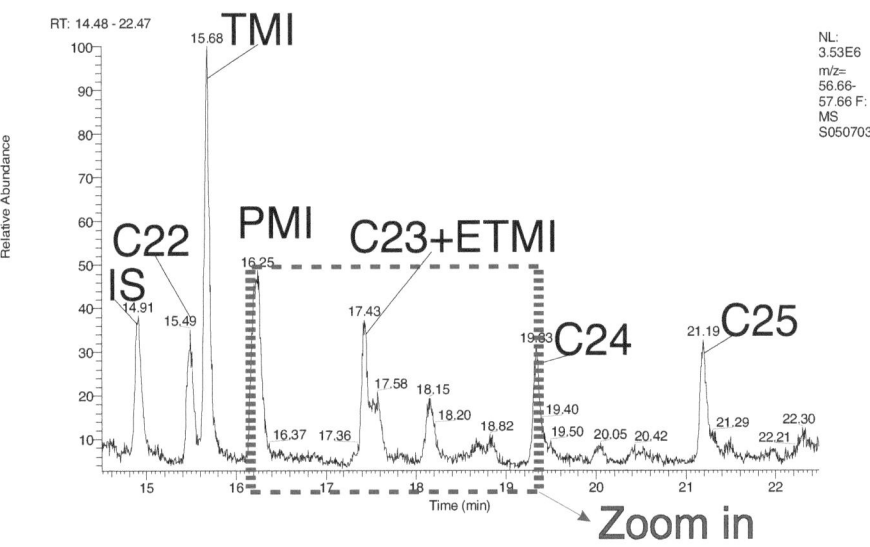

Fig. 44: Total ion current (m/z 57=n-alkanes) from sample VB124 (at depth 9,40 within the OAE interval). nC_{22}-nC_{25} (short chain n-alkanes), Internal Standard (IS) and the irregular isoprenoids TMI, PMI and ETMI. Zoom into Fig. 45.

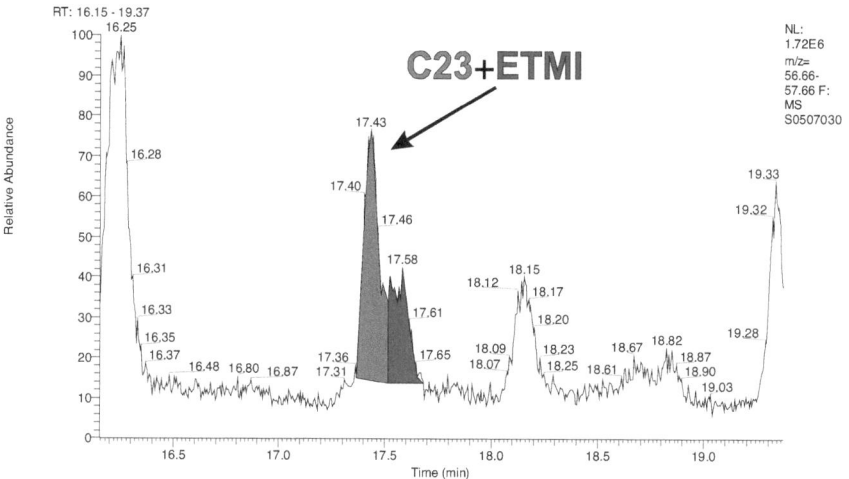

Fig. 45: Zoom into Fig. 44. Presented is the nC_{23} alkane co-elution with the ETMI isoprenoid displaying the arising difficulty in integrating precisely.

Same difficulties arise quantifying the abundance of the C_{40} tail-to-tail isoprenoid lycopane (2,6,10,14,19,23,27,31 –octamethyl dotriacontane; Schouten et al. 2000; Wagner et al. 2004; see also chapter 4.2.3.). It is a characteristic feature of the saturated HC fraction exclusively for the OAE interval as its occurrence is limited to the black shale. Analyses of the OAE samples did not achieve quantitative gas chromatographic separation of lycopane and the C_{35} n-alkane but the elevated concentration during the event is quantified as one combined peak (Fig. 46). Sinninghe Damsté et al. (2003) proposed to use the lycopane/C_{31} n-alkane ratio as a proxy to assess palaeo-oxicity during sediment deposition and showed that lycopane is indeed often abundant in sediments which were deposited under anoxic conditions. The conservative approach of Wagner et al. (2004) was followed here, quantifying the peak representing lycopane+ n-C_{35}, subtracting the contribution of n-C_{35} by assuming that it was present in 50% of the abundance of n-C_{33}, and calculating the resulting abundance of lycopane as modified lycopane*/C31 n-alkane ratio (e.g. Beckmann et al. 2008).

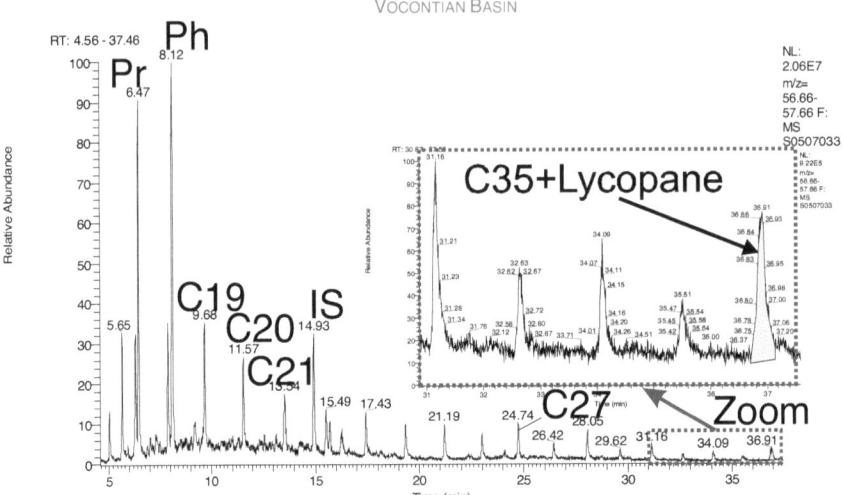

Fig. 46: A TIC of sample VB139 (at depth 9,72 m within the OAE 1b). The chemical composition is similar to sample VB124 (Fig. 44). Marked in yellow is the co-elution of n-C35 with the isoprenoid lycopane (at m/z 57).

As a biomarker for thermal maturity on a few sample selection the homohopane C_{32} αβ 22S/(22S+22R) ratio was calculated. The thermal maturity is an important factor for measuring the TEX_{86}. Values vary from 0.47 to 0.62 which is within the possible maximum. Values above 0.57 imply that the main phase of oil generation has been reached or surpassed. For measuring the TEX_{86} the data record should range from 0-0.2 denoting low sediment maturity. The high maturity of the presented data record unfortunately avoided a viable result. The assumption of highly thermal maturity for the presented data set is also supported by Vink (1996) and is not surprising regarding the high tectonic activities surrounding the Vocontian Basin.

All values for the Pristane/Phytane ratio lie around 1, which is rather low suggesting anoxic, commonly hypersaline or carbonate environments during OAE 1b. During the deposition of OAE, values are slightly <1, the remaining record imply values slightly >1 indicating rather suboxic deposition. Although this interpretation rounds quite well the here presented record off, the Pr/Ph-ratio is influenced by the thermal maturation of the sediments and is therefore not significant. Accordingly, it is not recommended that Pr/Ph ratios ranging from 0.8-3.0 are interpreted to indicate specific palaeoenvironmental conditions without corroborating data (Peters et al. 2005). Ratios above 3 might indicate terrigenous plant input deposited under oxic to suboxic conditions. Ratios lower 0.8 indicate saline to hypersaline conditions associated with evaporite and carbonate deposition. The high maturity of the VB-samples has led to a considerable loss of potential biomarkers. Therefore, these sediments are not ideal for making palaeoenvironmental reconstructions upon the organic composition.

Regarding the tectonic evolution at the Vocontian Basin, high maturity is not surprising. The first foreland basin formed during Late Cretaceous times in conjunction with the Pyrénées-Provençal tectonic phase. The second foreland basin developed in Late Eocene-Early Oligocene (Nummulitic Sea) during the second Alpine tectonic phase, and the third was present during the Miocene in connection with the late Alpine tectonic phase (Pittet 2008).

5.2.4. Inorganic composition

The composition of the clastic flux displays strongly cyclic variations in the order of small scale variations and probably representing precession controlled orbital forcing (Herrle 2002 and Herrle et al. 2003). These precession cycles (assuming ~20 kyr for each cycle) are most dominant in the Si/Al record (Fig. 47). Therefore, the Si/Al record and the Nutrient Index (NI, presented by Herrle 2002) were used together with the carbon isotopes to establish a time model for sedimentation rates for Site 545 (see chapter 4.2.5.). In the pre-OAE interval the Si/Al record swings very pronounced and steady within cyclic low amplitude variations from 2.5-3. Zr/Al parallels the Si/Al record, showing the same trends ranging from 13.8-17.4. Both element ratios are in significantly lower concentrations present than at Site 545. The SEFB was not primary influenced by coastal upwelling during the Cretaceous as proposed for Site 545 and not directly situated at an extensive land mass.

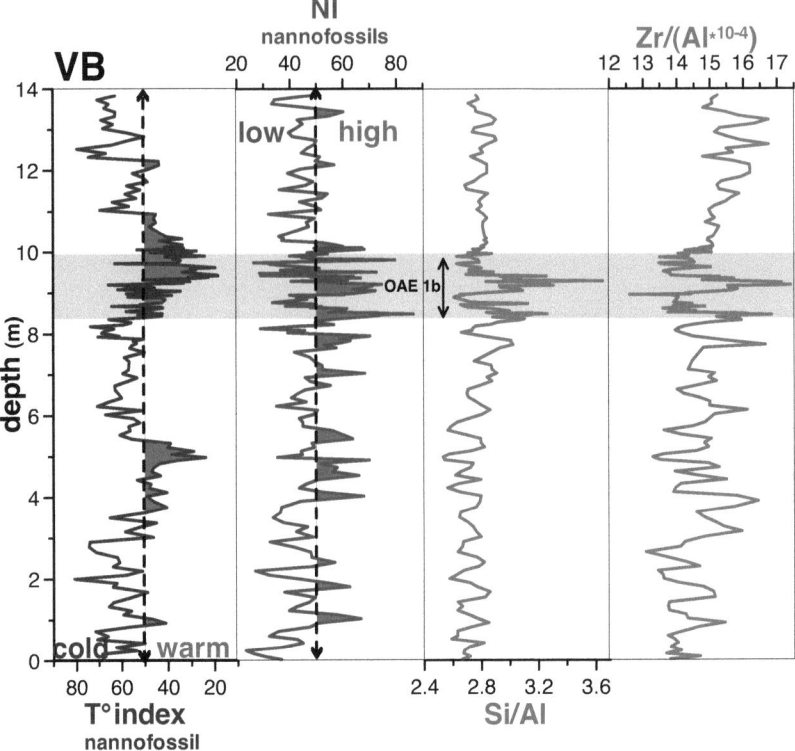

Fig. 47: Geochemical record from the Vocontian Basin presenting the inorganic composition. TI and NI based on calcareous nannofossils after Herrle (2002). High values of the nutrient index indicate elevated fertility and vice versa. Low values of the temperature index indicate high temperature and vice versa. Contrary to Site 545, ratios remain constantly fluctuating during the OAE 1b interval.

Both Si and Ti and Si and Zr correlate significantly (Fig. 48) which strongly indicates a common terrestrial source for the three elements (Hofmann et al. 2008). Embedded in drowned and condensed carbonate horizons, displaying the temporal evolution of the distal portion of the northern Tethyan carbonate platform during the Lower Albian, the deposition of OAE 1b is accompanied by a influx change into a mainly siliciclastic fraction delivered from the middle and south Helvetic realm (Föllmi et al. 2006). Changes in the influx occur at the same time when carbonate platforms begin to drown. The deposition of siliciclastic material is also confirmed by the recognition of three main subsidence phases

(Valangenian to Albian age) and their fraction dominated sediments (Wilpshaar et al. 1997; Arnaud & Lemoine 1993).

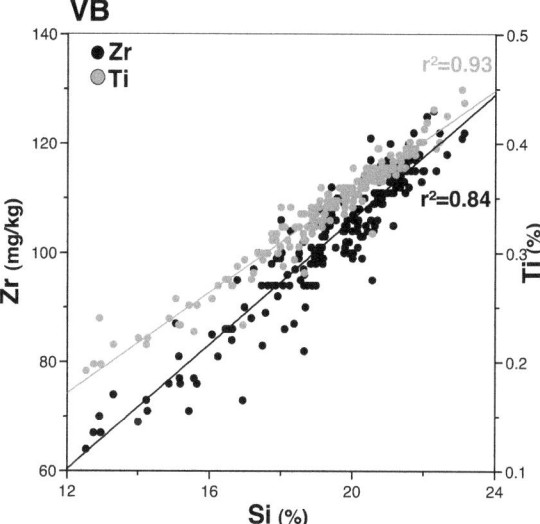

Fig. 48: Correlation of Zr, Ti, and Si. The good correlation suggests a strongly common terrestrial source for these elements similar to Site 545.

In order to obtain a multivariate and visual analysis of the mineral composition, the molar proportions of Al_2O_3 (A in Fig. 49), Na_2O+K_2O (NK of Fig. 49, A) and Fe_2O_3+MgO (FM of Fig. 49, A) are portrayed in a A-NK-FM ternary plot (Nesbitt & Young 1996). The CaO component was canceled because per definition it may only reflect the amount of CaO incorporated in the silicate fraction of the rocks but the described samples possess a relative high amount of carbonate concentrations. The samples show high concentrations of aluminium (in the A-N-K diagram, Fig. 49, B) displaying highly weathered material which is also confirmed by the CIA (see below). Quartz is the only major mineral of the sediments that does not plot in a A-NK-FM diagram. Apparently, the availability of aluminium is linked to the delivery of Si and Zr into the system, inferential aluminium is present in high amounts in the A-N-K ternary plot. All samples plot in a narrow field close to the A-Apex in both ternary plots and above the feldspar field (on the left-hand boundary at 50% Al_2O_3). Samples for the pre- and post-OAE time period are almost of the same chemical composition overlaping each other and only samples for the OAE-interval extent slightly these limits.

The composition of the samples is very homogeneous, without clear trend lines for pre-, during and after OAE intervals. But while the composition before and after the event is very similar, samples from the OAE interval show a slightly higher diversity (see Fig. 49, A). Furthermore since all the samples plot near the Al_2O_3 –Apex, it is assumed that sediments from the Vocontian Basin contain high amounts of clay minerals (Nesbitt &Young 1996). In the A-N-K diagram, Illite and Smectite would plot at 70-85% Al_2O_3 and the other major clay mineral groups (kaolin, chlorite, gibbsite) plot at the A-apex (100% Al_2O_3).

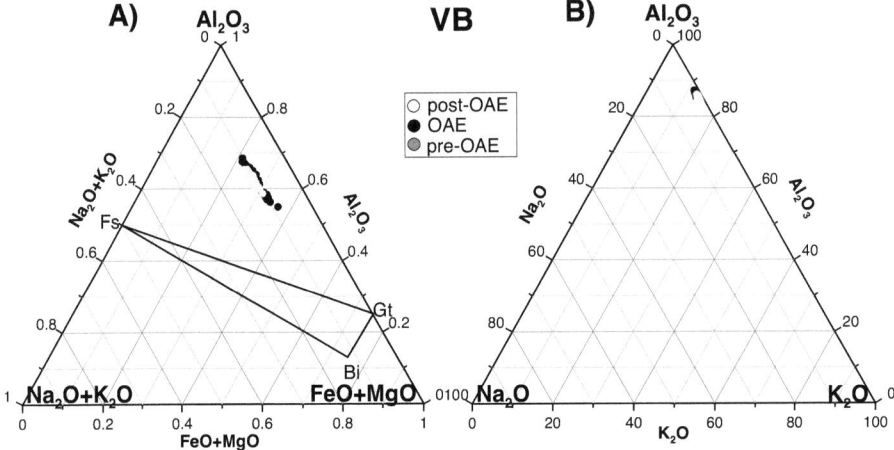

Fig. 49: A) A-NK-FM ternary plot and B) A-N-K diagram (after Nesbitt &Young 1996). All samples plot very narrow without clear distinctions and reveal high Al concentrations.

The different parameter profiles show clearly that the cyclic fluctuations begin prior to the OAE and strikingly remain stable during the event with an enhanced amplitude, demonstrating stronger contrasts. Obviously all analysed major elements reflect a disturbance of the cyclic variations after the event, probably due to the tumbling OAE conditions. Compared to the duration of OAE itself (~45 kyr, see Fig. 28 and next chapter) the time of disturbance after the event requires approximately the same time period (~44 kyr, see Fig. 50).

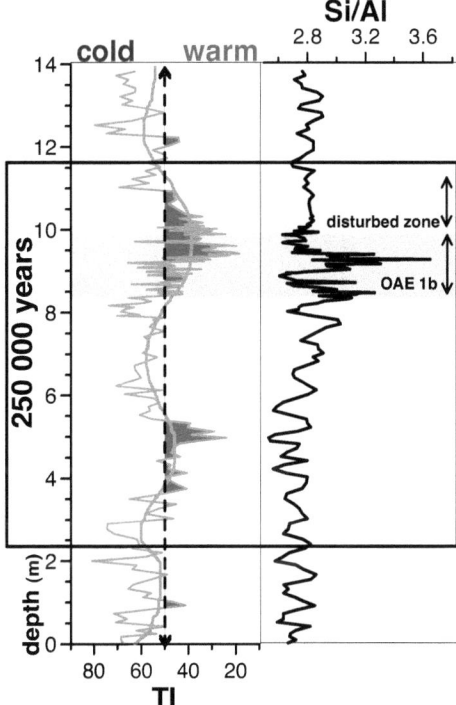

Fig. 50: Cyclic variations displayed by the TI and the Si/Al ratio were the base for the time-model (see 5.2.5.). The disturbed interval indicated by the constant Si/Al ratio after OAE 1b is approximately as long as the black shale interval itself.

After this lack of time, cyclicity is re-established for the remaining 2,5 metres of the record. In contrast to Site 545, the element ratios of Si/Al, Zr/Al and K/Al display no increase at the onset of OAE 1b but rather a decrease (Fig. 47 and Fig. 51). The clastic flux is displayed by the elements Al, Si and Zr. Na+Ca and K are elements resulting from feldspar weathering or, in the case of calcium, essentially in biological processes and transported in general fluviatile. Accumulation rates of Al, Si and K are uniformly. During the OAE, ratios decrease but still reveal constant fluctuations showing stronger amplitudes. The ratio of Na+Ca reflects in contrast counter-cyclical behaviour. High ratios go confirm with lowest ratios presented by the remaining clastic flux parameters. In the uppermost part of the OAE a zone of interruption is visible where the records of all discussed parameters parallels each other. This zone commences in the disturbed interval after the black shale event before `normal´ conditions are re-established at the top of the record.

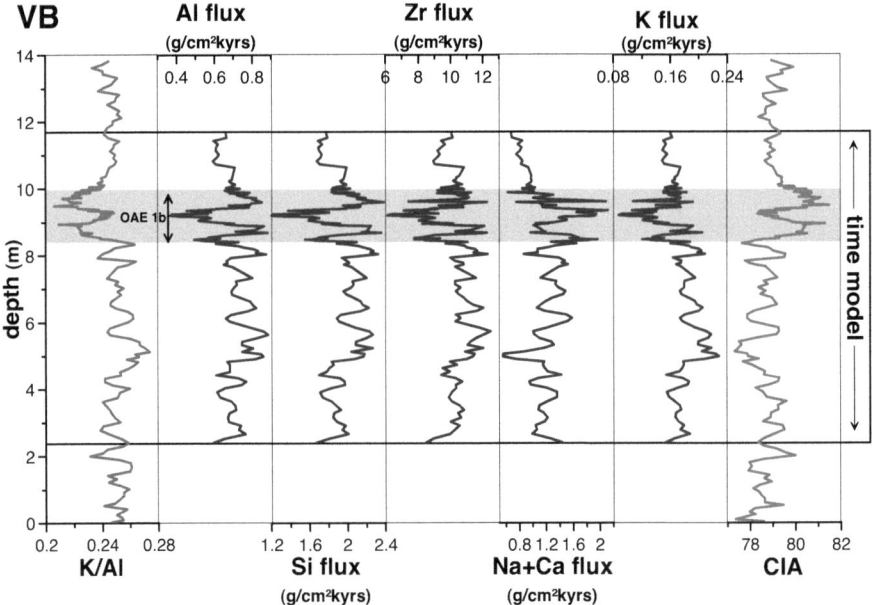

Fig. 51: Shown are the K/Al ratio, flux rates for Al, Si, Zr and Na+Ca, and the Chemical Index of Alteration (CIA). Note, flux rates are measured after the time model (see 4.2.5. and 5.2.5.) and reflect a limited sample record.

The CIA at the Vocontian Basin (Fig. 51) presents values in the range between 77-81 and are therefore slightly higher than at Site 545 (CIA= 72-75). During the OAE 1b CIA-values rise towards the top of the event, still subject to preccessional variations. After highest values around 81 they decrease stepwise towards the ending of the black shale and display the same disturbed phase after the event as already discussed. A higher degree of weathering for the studied samples is suggested and the presence of clay minerals was supported by the A-N-K and A-NK-FM ternary plots.

Like at the Mazagan Plateau the redox-sensitive trace metals (TM) presented for the Vocontian Basin are Ni, V and Zn (as minor elements in ratio to Al $*10^{-4}$). With respect to average shale values (AVS; Brumsack 2006; Fig. 52) TM are mainly enriched during the black shale event. Particularly the V/Al ratios are high within the interval at values between 20 and 54 and ranging also in the remaining record slightly above the AVS with values between 12 and 20. During the black shale, highest values are reached when the TOC content was lowest. Ni/Al ratios before OAE 1b fall slightly above AVS data with values fluctuating between 9-11. Values triple at the beginning of the black shale but decrease subsequently to a range between 7.7. and 14 still within the OAE. Post-OAE data are at first lowered, than rising again to pre-OAE values in the upper part of the record. Zn/Al ratios are enriched and with values between 13-33 before and after the event up to three times higher than AVS values. Unexpectedly, lowest values below the AVS data are measured during OAE 1b. A few positive peaks exceeding values of 40 are present but negative values maintain stable until the post-OAE stage begins. Positive values during OAE 1b go up to 54.

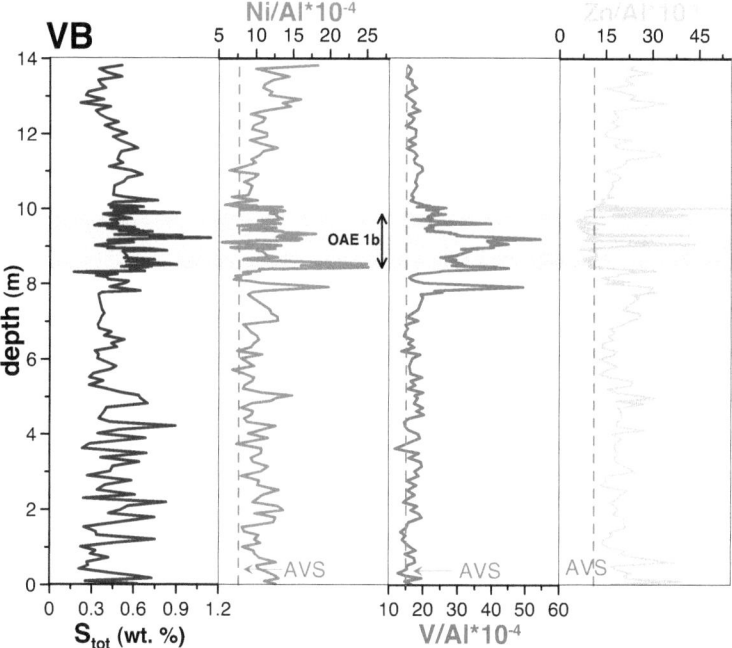

Fig. 52: Variation of selected redox-sensitive elements and element ratios for the Vocontian Basin. The broken lines represent average shale values (AVS from Brumsack 2006). Note that the OAE 1b interval is not significantly enriched in redox-sensitive elements and, in the case of Zn, it is even depleted.

Trace element concentrations are higher than at Site 545 but in comparison to other Lower Albian black shale Sites still low (see Chapter 4.2.4. for details). After Brumsack (2006) the data presented here would correlate best with Unit 1 data from the Black Sea, except the elevated Zn/Al ratio (Black Sea, Unit 1=18) which is in any case not easy to interpret showing lowest and highest values during the black shale. The black Sea is interpreted as a semi-enclosed basin reflecting euxinic conditions (Tribovillard et al. 2006). The TM Vanadium is seen as a redox proxy, present in an enzyme used by nitrogen-fixing bacteria, with minimum detrital influence whereas Nickel and Zinc are seen as productivity proxies and behave usually primary as micronutrients. The enrichment of V during the OAE 1b interval in the presence of sedimentary OM (Ni enrichment) suggests reducing conditions similar to the Black Sea although euxinic conditions were absent. During the formation of OAE 1b a lowered detrital influx can be suggested. There is no correlation of Ni and Zn concentrations neither with TOC content nor with total sulphur content (r^2 for both elements and TM under 1.5, Fig. 53) but the palaeogeographical situation for the VB reveals difficult relationships displaying a semi-restricted marine basin. Both the marine circulation pattern as well as delivery by carbonate platforms were important for the sediment supply, displayed by the cyclic variations.

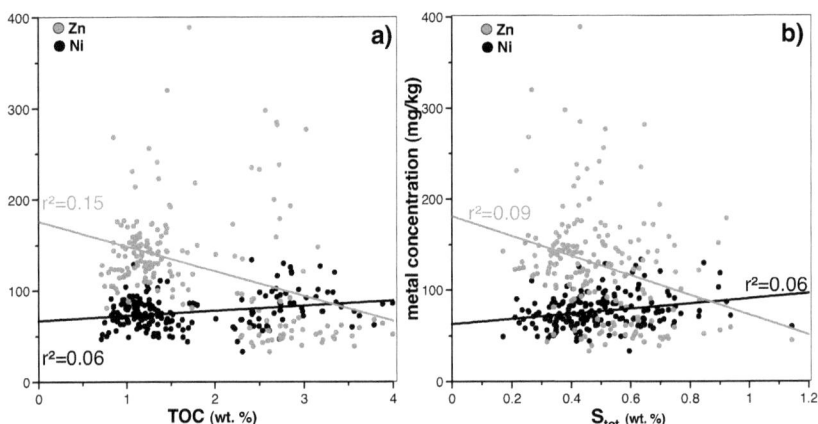

Fig. 53: (a) Correlation of Ni, Zn and TOC and Ni, Zn and S_{tot} (b) at the Vocontian Basin. No correlation leads to the suggestion that Ni and Zn did not act primary as micronutrients.

It is evident that also at the Vocontian Basin anoxic water conditions were absent. In view of the benthic foraminifera repopulation events in the OAE 1b interval (Friedrich et al. 2005) and the Vanadium enrichment (Tribovillard et al. 2006) lower oxygen levels are assumed, probably reflecting suboxic conditions maybe becoming almost short-termed anoxic.

5.2.5. Time model and sedimentation rates

All main characteristics for the time model are already explained in detail in Chapter 4.2.5. for the Mazagan Plateau. The base for the presented time model were cyclic patterns described as precessional fluctuations in the Vocontian Basin record, characteristic shifts in the $\delta^{13}C_{carb}$ curves for both sites and the assumption of 45 kyr duration for OAE 1b. This construct simplifies the synthesis when both sites are compared. At the Vocontian Basin highest sedimentation rates (on average 3.7 cm/ka) occur before and during OAE 1b and are after the event lower, but still higher than maximum SR for the Mazagan Plateau. In total amounts, SR are not higher during the event as prior to it and rates even drop in the middle part of OAE to increase again and decrease towards the ending of the black shale sedimentation (Fig. 54). Post-OAE SR are first as low as earliest rates before the OAE sedimentation but still decreasing to the lowest rates of the record (3.0 cm/kyr). The entire record presents according to LSR a duration of 250 kyr and OAE 1b formation occurred within ~45 kyr. The time model represents, due to the characteristic shifts in the $\delta^{13}C_{carb}$ curves (see Fig. 28), only an abstract of the entire record which extends in both directions (Fig. 55). Because of this fact, most parameters are presented in concentrations or ratios to depth assuring highest sample resolution.

As well as for Site 545, to display the development of the siliciclastic flux and organic matter accumulation, the flux rates of Al, Si, K and OC as selected element accumulation rates are presented (Fig. 51 & 54). The results are as expected strikingly different than at Site 545. Prior to the OAE sedimentation, the organic carbon (OC) flux reveals fluctuating de- and increasing concentrations at values between 0.08–0.31 g/cm^2/yr, where elevated values occur apparently due to the other OM-rich layers (see Chapter 2.1.). At the base of the event the increase of the OC flux is rapid and remain at constant elevated values during the entire event around 0.2-0.4 g/cm^2/yr. Within this range short scale fluctuations are obvious and dominant. The constantly elevated OC flux is shortly interrupted by a lowered accumulation of OC around 0.2 g/cm^2/yr. This negative peak goes confirm with the drop in the SR during the OAE 1b and the other parameters presented here.

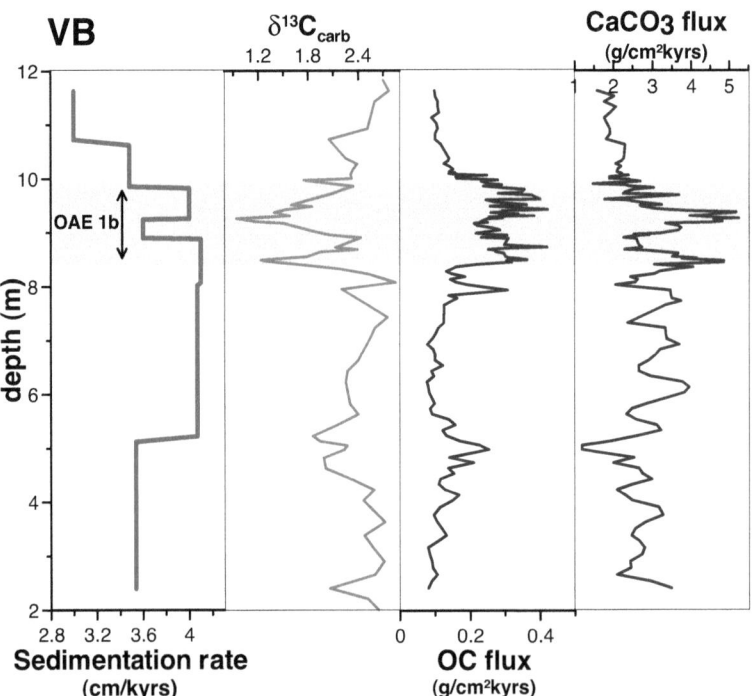

Fig. 54: Sedimentation rates, organic carbon- (OC) and carbonate carbon (CaCO$_3$) flux rates presented for the Vocontian Basin in comparison to the bulk carbon isotopes.

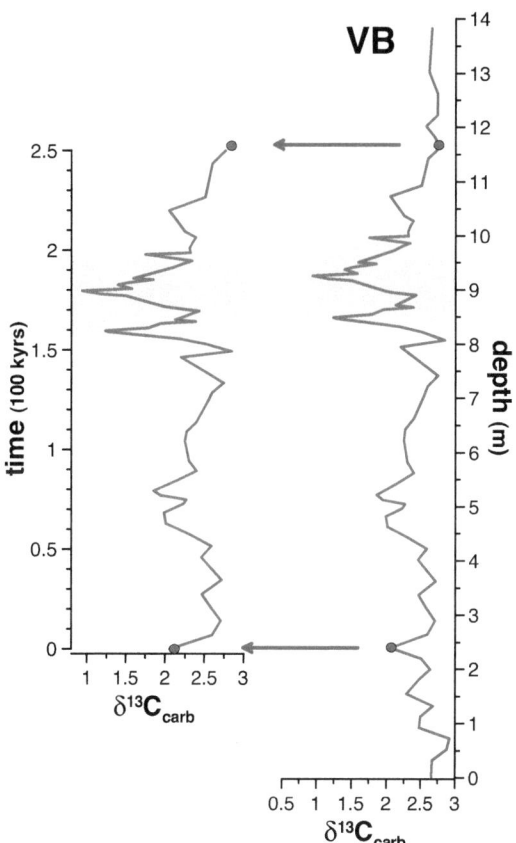

Fig. 55: Left, the time model (comprising 250 kyrs) resulting from characteristic shifts in the isotopes (Fig. 28) and presenting thus only a limited interval of the complete data record (in metres), shown at the right.

Afterwards values increase again and decrease stepwise in the upper 26 cm of the black shale. The OC flux in the post-OAE record remains low, still indicating a decreasing trend. Strikingly, although values decrease during the drop in SR within the event, rates of OC flux are almost doubled as comparable rates before the OAE. OC burial rates are higher than at Site 545 but still not as high to support a major carbon burial event.

The accumulation rates of Al, Si and K match with the SR. Constantly fluctuating and showing a slight increase when SR increase prior to the event (to 4.07 g/cm^2/yr). During the event constant fluctuations as well as the sharp drop in accumulation rates in the middle part are distinct. Except potassium, rates obtain again highest values and decrease towards the top of OAE 1b. Potassium behave in the first half of the event similar to Al and Si. In the second half when SR increase again after the drop, lowest K rates go parallel to highest Si and Al values as a single peak but record the same decrease in the uppermost OAE as mentioned before. The post-OAE stage show decreasing rates parallel to the stepwise decrease of the SR. Apparently, neither Al nor Si are enriched elements during the event contrary to accumulation rates at Site 545. The ratio of Si/Al display a generally balanced system where Si is slightly enhanced during the OAE compared to the remaining record (Fig. 47). Results suggest, that apparently detrital continental derived nutrients were not as important for higher OC-accumulation rates as it is indicated for the Mazagan Plateau or that primary productivity (PP) was rather stable and preservation conditions increased.

5.3. Discussion

On a geological record from the Vocontian Basin continuative geochemical analyses were performed to gain as much information as possible. With regard to the millennial scale record provided for the Mazagan Plateau and the promising results, same procedures and analytical efforts were obtained on the VB data record. Although consistent comparison was impossible due to the different composition and maturity grade of the analysed matter, the results provide new insight of the complex relationship between atmospheric and biological processes due to black shale sedimentation. At the VB the `abrupt´ temperature rise in sea surface documented for Site 545 was attempted to recognise by highlighting the cooperating factors as nutrient delivery, excess OM burial and the resulting disturbance of a `normal´ functioning ecosystem. Unfortunately, TEX_{86} measurements were not conducted due to higher maturity of the samples. Nevertheless, one striking result is that the VB represents a totally different palaeoenvironment displayed by various inorganic element-ratios albeit the palaeogegraphical position and their adjacency to each other. For the first time the terrigeneous runoff was identified in detail and compared to the OM. Orbital forcing is visible and dominant for the entire record and persist through the black shale formation. The presented results are in the following discussed in detail including climatic, productivity and tectonic states. A time schedule for OAE 1b formation and the disturbance of an eco-system will be discussed, what triggered black shale formation and what has changed during its formation. Predictions about the climate in this certain environment, fluctuations in the bioproductivity cycle and the associated origin of nutrients and delivery pathways are possible. As introduced for Site 545, the presented and already shortly described results from the VB will be discussed, divided into three sub-chapters grouping the discussion into a pre-OAE, a OAE and a post-OAE stage. This sub-division should highlight the changes ongoing in a complex climate system during different phases and simplifies explanation and comparison of both sites in the Synthesis Chapter.

5.3.1. Conditions prior to OAE 1b

Sedimentation prior to OAE 1b was clearly dominated by precessional orbital forcing. These small scale variations in the order of 20 kyrs are recognised in almost every analysed parameter and were the base of the time model construct presented in the previous chapters. The Temperature Index (TI) based on calcareous nannofossils (Herrle 2002) reflect these smaller variations but show ostensible a predominant variation in an superior orbital forcing, comparable to the proposed eccentricity controlled system at the Mazagan Plateau (Fig. 56).

Although temperatures can not be stated in degree Celsius as given by the TEX_{86}, a broad declaration into warmer and cooler intervals as well as general climate trends can be made. One warm-cold cycle is defined as described earlier for the Mazagan Plateau: from a beginning warmer state to the ending cooler state. The first temperature maximum prior to OAE 1b is accompanied by the second organic rich shale deposition at the Vocontian Basin (HN8; 4.73-4.98m) which is also documented in the TOC- and $CaCO_3$ content (Fig. 38). The first shale deposition (HN7; 0.33-0.60m) is not pronounced but goes along with a beginning warming cycle which than declines into a cooler interval.

The record starts with a warm interval which soon declines into a cooler state, followed by a complete warm-cold cycle and the proximate beginning of a warm phase. Suggesting the warm-cold fluctuations from the TI reflect the frequency of eccentricity cycles and the small scale variations from the Si/Al ratio precessional cycles, the record prior to the black shale event represents approximately 200 kyrs (Fig. 50). This assumption is in good correlation with the time model based on the carbon isotope profile, presenting 150 kyrs for the shortened interval (see chapter 5.2.5. for details).

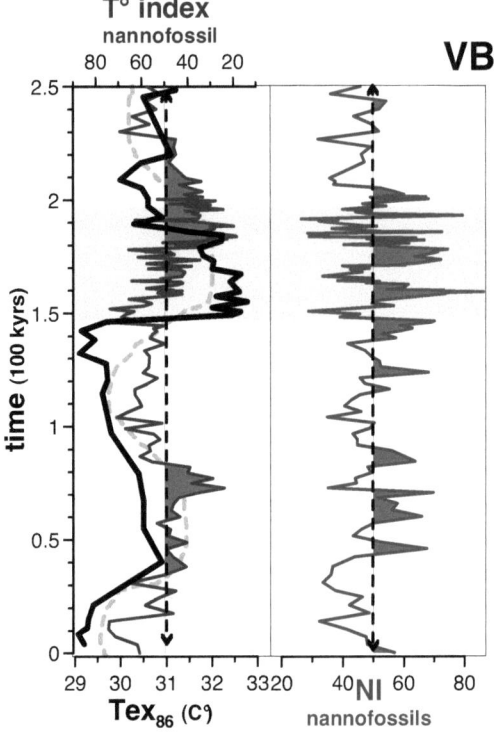

Fig. 56: Comparison of the Tex_{86} (taken from Site 545) and the temperature index, nutrient index (for the Vocontian Basin; TI, NI after Herrle 2002). The dashed grey line reflects a simplified temperature trend and reveals thus similarities for both sections.

The temperature curve is apparently controlled by the higher ranked eccentricity cycle (Herrle 2002). Obviously, the sedimentation into the Vocontian Basin was controlled by precessional orbital forcing regarding the remaining parameters during pre-OAE times (and the Nutrient Index by Herrle 2002). Contrary to the MP the record reflects mainly the preccesional signal with smaller scale fluctuations in repeating periodic patterns. A superior cycle as demonstrated for the temperature curve is not observed. Contrary to all parameters the organic carbon accumulation presents in general a stable almost non-fluctuating record without board variations with the exceptions of the HN8 black shale sedimentation and a peak pre-dating the onset of the OAE 1b shortly. Highest values occur during the HN8 shale formation which parallels the high-stand of a warm interval. During this highest temperatures in pre-OAE times, the bio-system was not disturbed as indicated by the remaining parameters. But the elevated rate of carbon burial is documented in the carbonate record and in the carbon isotope profiles. Obviously the surplus production of organic carbon was a short-lived event during warmest temperatures without enduring or permanent effects on the bio- and climate system.

The discussed superior warm-cold cycles prior to the OAE 1b are otherwise not visible in the remaining parameters displaying the inorganic composition of the samples. Here, all parameters reveal a stable record with periodic fluctuations in the range of precessional cycles, which is also recorded in the NI. During times of higher fertility, as example the two black shales predating the OAE, the TOC is high and at the same time the amount of terrestrial derived matter (ratios of Si, Zr, and K relative to Al) are low. Herrle (2002) postulated a strong monsoonal circulation in the Vocontian Basin area, where changes are associated with strong fluctuations in wind velocities and precipitation rates. Apart from that assumption, the accumulation of OM at the Vocoantian Basin was definitely highest during humid periods when wind velocities were probably weakened but fluviatile runoff was elevated due to increased precipitation rates. Furthermore the element ratios of Si/Al and Zr/Al are only present in low concentrations suggesting a lower importance of the detrital fraction amount compared to Site 545. During arid and potentially cooler climate intervals deep water formation was probably enhanced (Herrle 2002). This could have triggered the mixing of deep water masses favouring ocean currents probably due to changing wind directions.

During Lower Albian times the position of the Vocontian Basin reflects a semi-enclosed basin. Although, the Tethyan to the south-east was not completely wide opened as assumed (Fig. 57). Apulia was situated in the SE almost in front of the Vocontian Basin and apart from that, the ocean pathways within the Mediteranean Sea were not really open but restricted due to various continental terrains. Reconstructing all wind directions and ocean currents for this time interval is therefore not possible but reveal a stronger focus because they highlight climatic consequences. Transport ways and the composition of the organic matter and sediments depended strongly on palaeoclimatic, -tectonic and -oceanographic changes. As discussed by the A-N-K ternary plot (see chapter 5.2.4.) the content of Al_2O_3 in all samples lies between 80-90% suggesting that clay minerals were the most abundant minerals.

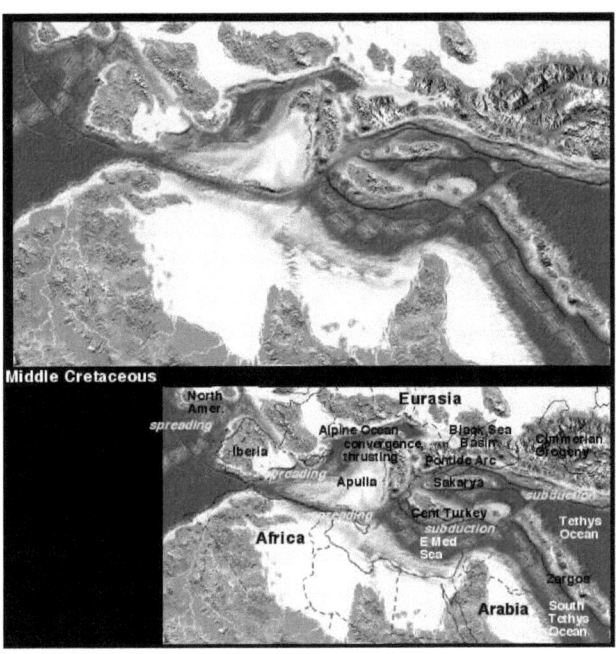

Fig. 57: Sedimentation, Tectonics, and Palaeogeography of Southern Europe and the Mediterranean Region (after Scotese 1998). The top map shows the un-labeled palaeogeography and the second map labels the tectonic and palaeogeographic elements (from: http://jan.ucc.nau.edu/~rcb7/midcretmed.jpg).

Chemical weathering at the Vocontian Basin was significant not at last due to the open marine palaeogeographical setting. Transport ways, until the sediment fraction finally arrives at the investigated section, were enormous and favoured extensive weathering of the sediment fraction. The CIA is therefore slightly higher at the Vocontian Basin and also due to the geographical situation fluviatil delivered particles might have been the dominant way of sediment supply. This sediment basin was exposed to extreme climatic changes, displayed by the precessional cycles and assumed by the monsoonal theory (Herrle 2002). The terrigenous input delivered to the Vocontian Basin was preferentially transported by runoff and ocean currents while major wind systems probably played a minor role. Over the time period of nearly 200 kyrs the bio-system at the Vocontian Basin reveals stable, strongly affected by precessional fluctuations but without exceptional excursions. Maximum Mg/Al ratios occur usually (except during OAE 1b) at the end of the cooler intervals before they decrease in the warmer intervals (Fig. 58). The climatic situation was most likely comparable to the MP with humid conditions during warmer intervals and colder, dryer climate intervals forcing deep water formations.

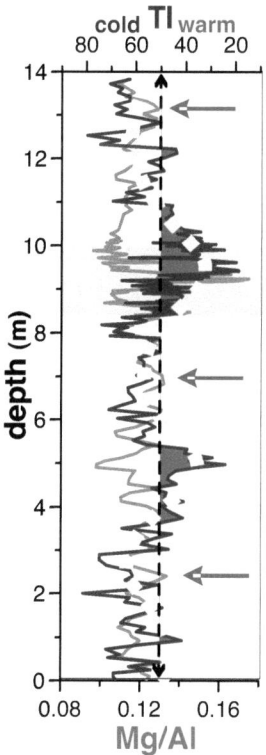

Fig. 58: Mg/Al ratio compared to the TI. The thick dashed line presents simplified general temperature trends. Arrows point to maximum Mg/Al ratios (except during the OAE interval) at the end of cooler intervals.

Most striking difference is the, at last nearly, positive water mass balance at the Vocontian Basin during humid intervals leading to approximated precipitation- to evaporation rates in contrast to the negative mass balance at the Mazagan Plateau where evaporation exceeds clearly precipitation (see Synthesis Chapter).

With TOC values around 0.7-2.8 % the organic carbon content for the record is in general low prior to the OAE 1b. The crossplot of the Zr concentration versus the TOC content is used as a measure for grain-size and wind strength (Fig. 59).

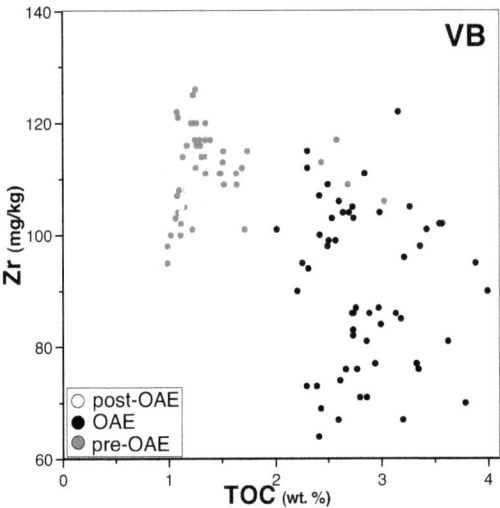

Fig. 59: TOC vs Zr crossplot for the pre- and post- OAE values show no linear correlation. Changing Zr concentration does not affect the range of TOC values. OAE samples show higher TOC values during low Zr concentrations.

Contrary to Site 545, the record displays a trend of higher TOC values with decreasing or constant Zr values in the pre-OAE section. This confirms the postulated lower importance of wind systems at the VB and supports higher runoff transported by river flow and nutrients delivered by deep water formations. As discussed (see chapter 5.2.2.) the organic matter composition presents a broad field for origin and sources. Nevertheless, a clear distinction is suitable as samples prior to OAE 1b present HI around 100 mg HC/g TOC with the exception of the remaining black shale layers. Here values are slightly elevated with 135-164 mg HC/g TOC, postulating a mixture of marine and terrestrial derived organic matter. The remaining record reveals clearly that the origin of the organic matter is derived from terrestrial sources. This is also supported by the presence of long-chain odd-numbered alkanes through the entire interval which is confirmed by compound-specific carbon isotope measurements. As marker of methanogenic activity the irregular C_{24-26} isoprenoid skeletons TMI, PMI and ETMI are found during the black shale formation and its bordering sediments. They first appear around 50 centimeters prior to OAE 1b but during a black shale sedimentation (at depth 7.93m) suggesting an expansion of the assumed OAE sedimentation window. CPI values are below 1.0. As there are no indications for hypersalinity and/or water column stratification (Vink 1996) and the Vocontian Basin was also surrounded by drowned carbonate platforms during the Lower Albian, this might also point rather to a carbonate environment than hypersalinity. A crossplot of the S2 pyrolysis data and the organic carbon content (Fig. 60) reveals clearly the changing composition of the OM from pre- as well as post-OAE stages to the OAE interval and indicates rather a temporal change in major nutrients delivery pathways or points to inceptive better preservation conditions.

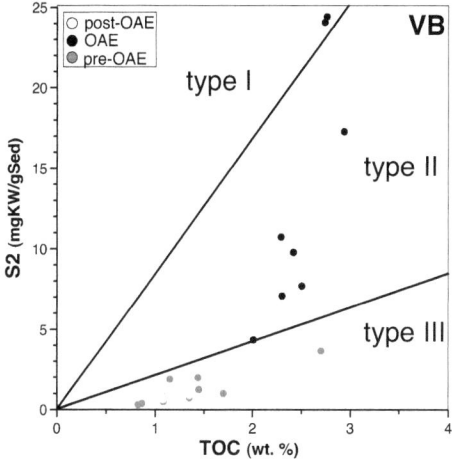

Fig. 60: S2 versus TOC crossplot. Boundaries for Types I, II, and III kerogen fields are taken from Langford and Blanc-Valleron (1990).

Prior to OAE 1b the TOC content does not correlate linear to the detrital siliciclastic fraction (presented by the Si/Al ratio, Fig. 61) and values are, compared to the Mazagan Plareau, relatively low. On contrary, the concentration of aluminium in the sediment is almost twice as high as for samples from Site 545 but with nearly the same TOC concentrations (Fig. 62). This also emphasises the assumption of high precipitation rates, intensified fluviatile runoff, marine transport ways, abundant clay mineral assemblages and to a lower amount aeolian sediment transport. In pre-OAE times the iron content was elevated while the TOC content was lowered (Fig. 42).

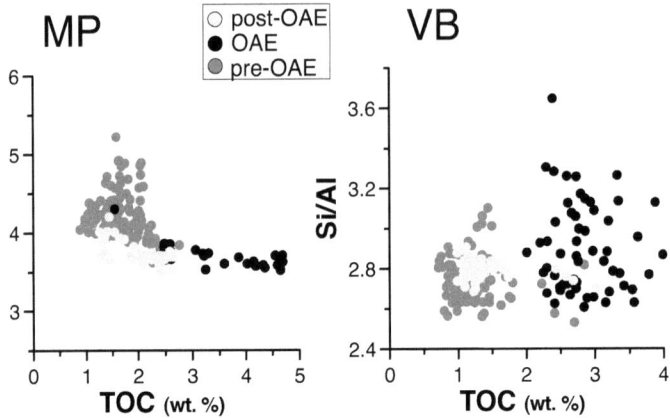

Fig. 61: Si/Al ratio versus TOC concentration displayed for both sampled sections. On the left Mazagan Plateau (MP) and on the right the Vocontian Basin (VB). Note, y-axes vary in scale. Higher Si/Al ratios at the MP suggest stronger wind systems than at the VB.

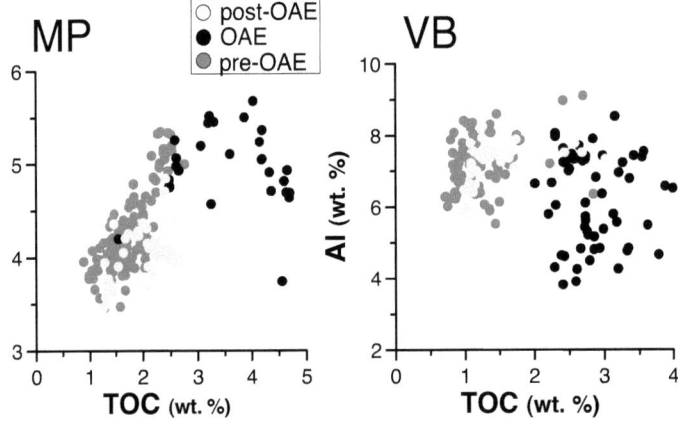

Fig. 62: Al versus TOC crossplot for both investigated sections. Annotation same as in Fig. 61. Note, y-axes vary in scale. Aluminium is enriched at the VB and does not primary depend on the TOC concentration.

As suggested for the Mazagan Plateau, low sulphur concentrations (around 0.5 %) may reflect an environment with moderate preservation conditions and low amounts of oxygen in the water column although iron limitation for the fixation of sulphur in the sediments is assumed. Anoxic water conditions were absent (see chapter 5.2.4.) but the assumption of lowered oxygen levels and rather disoxic conditions were supported by an enrichment of the TM vanadium and benthic foraminifera repopulation events in the OAE 1b interval (Friedrich et al. 2005).

The Vocontian Basin is characterised by precessional orbital forcing visible in almost all parameters. This analytical record provides a division into two states:

during **State 1)** higher siliciclastic fraction (Si, Zr, Ti) is coupled to higher carbonate productivity (Fig. 63, A).
during **State 2)** higher Al and elevated organic carbon content is observed when carbonate production and siliciclastic input was low (Fig. 63, B).

With regard to previous sequence stratigraphic records from the Vocontian Basin (Ferry et al. 1989) the general long-term sedimentation cycle was dominated by sea level variations (Fig. 64). During State 1) a lower sea-level stand would have favoured expanded sediment transport into the basin. Low sea-level stand was coupled to arid and probably cooler periods, when wind strength and evaporation were elevated resulting in a negative water-mass-balance (see Synthesis). A weak western or rather south-western wind system (Poulsen et al. 1999) could have transported the siliciclastic fraction into the deeper part of the basin without evoking coastal upwelling. Also possible are enhanced south-eastern winds leading to lagoonal currents during high evaporation and transporting the terrrigenous fraction from e.g. Apulia. This could have been possible in times when sea-levels were low and Apulia, made up of continental crust, was at least temporarily not water covered. Nevertheless, sedimentation into the Vocontian Basin is clearly approved and one of these thesis may be assumed but it is difficult to resolve from what direction and how material finally came into the Vocontian Basin. State 2) would imply high sea-level stand and reduced wind velocities due to warm and humid periods when mass balance is nearly positive and precipitation is elevated compared to evaporation. Enhanced fluviatile runoff and higher amounts of nutrients lead to enhanced PP. During this scenario, mixing of deeper water masses and re-oxygenation of the water column could also have fostered PP and higher accumulations of organic matter. The composition of the VB varies clearly in the range of precessional forcing. What finally drove these cyclic variations will be discussed in detail in Chapter 6.

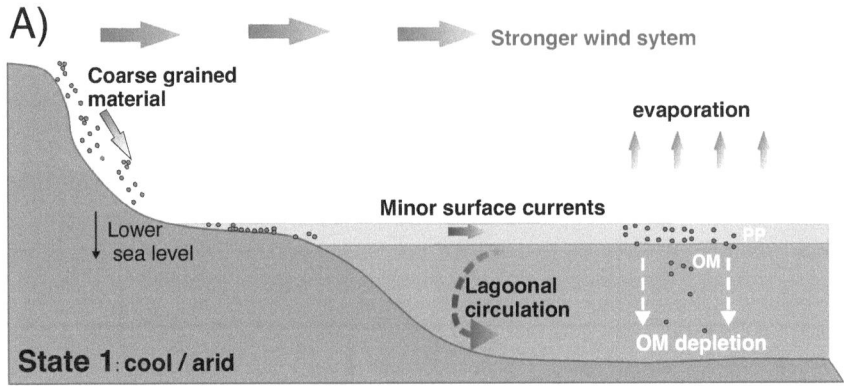

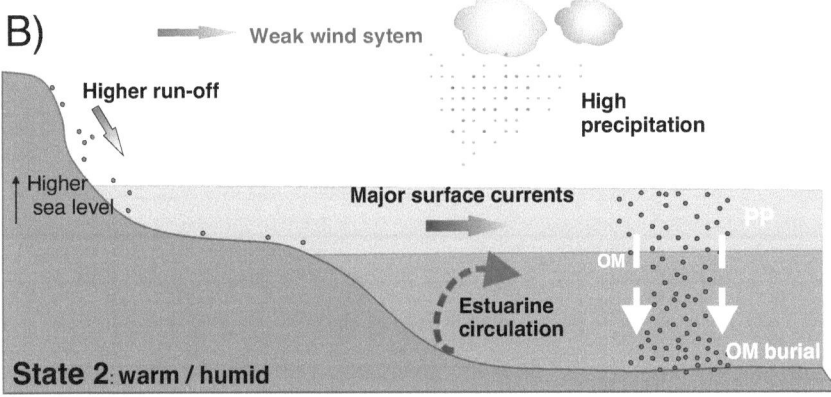

Fig. 63: Sedimentation model for the Vocontian Basin. During state 1 (**A**) the climate is cool and arid with stronger wind systems, high evaporation- and low precipitation rates, and high amounts of siliciclastic fraction. PP is reduced and OM burial is restricted. Climate conditions during state 2 (**B**) are warm and humid, the sea level is higher, higher runoff resulting from higher precipitation rates, major surface currents transporting Al and OC into the deeper part of the basin stimulating PP and providing OM burial.

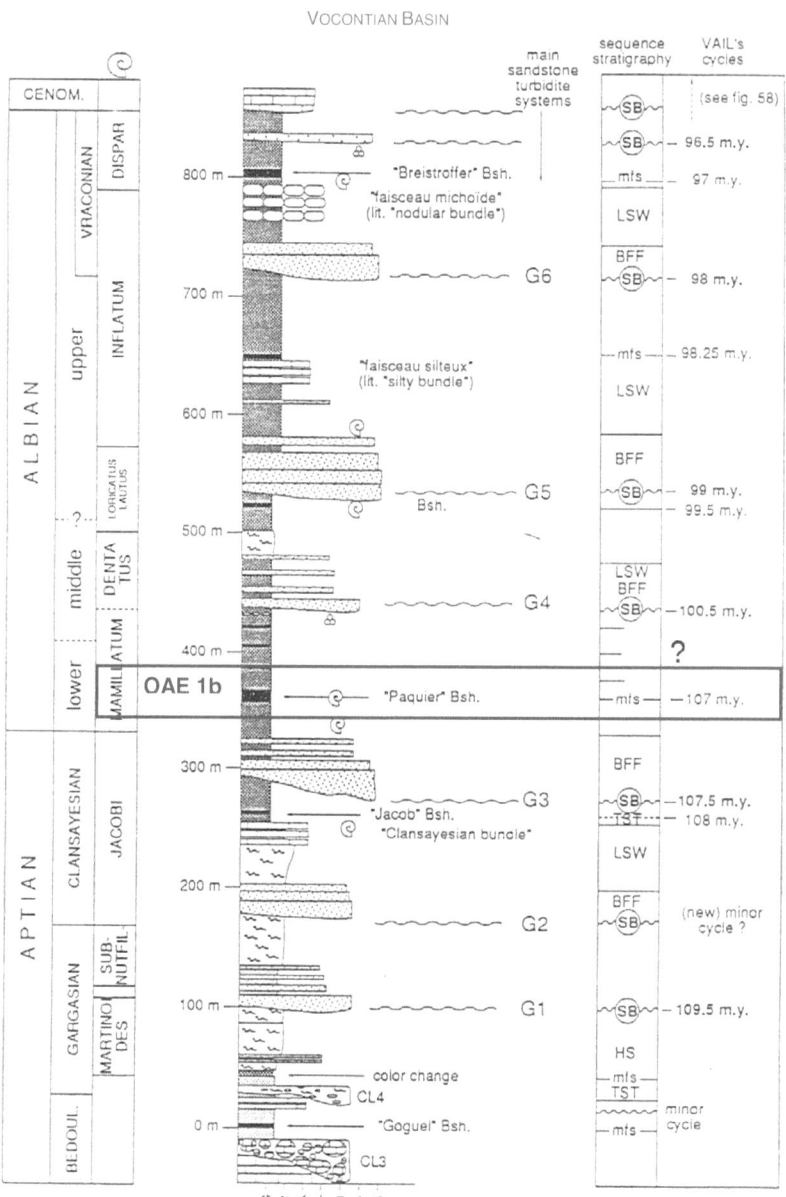

Fig. 64: Sequence stratigraphic record of the VB. Virtual basinal succession of Aptian-Albian marl deposits (after Ferry & Rubino 1989). The red rectangle highlights the OAE 1b interval.

HS=highstand; mfs=maximum flooding surface; SB=sequence boundary; BFF=basin floor fan; TST=transgressive system tract; HST=highstand system tract; LSW=lowstand wedge.

5.3.2. OAE 1b at the Vocontian Basin

The beginning of the OAE 1b burial is accompanied by a starting warm cycle. Obviously and very contrary to the Mazagan Plateau, all analysed parameters still display same cyclic fluctuations as prior to the event. The bio- and climate system seems at first sight nearly undisturbed. While almost all parameters including the carbon isotopes hardly mark a striking difference to the remaining record, the TOC content rises abruptly with the beginning OAE, yielding values almost three times higher as before (Fig. 54). Values remain elevated during the entire black shale interval with stronger fluctuations and drop again abruptly at the end of the interval to similar values as reported prior to the black shale sedimentation. The shift in the carbon isotopes is evident but it is not as prominent as at Site 545. For the Vocontian Basin, shifts with almost similar magnitudes are observed earlier during the previous black shale events, still the OAE 1b is marked by a stronger shift to negative values which is unique in the entire sample record.

OAE 1b sedimentation is accompanied by rising temperatures and enriched nutrient delivery (see TI, NI; Fig. 47). The OAE interval starts with the beginning of an warm/cold cycle reaching the warmest period during the upper third and revealing a decreasing temperature trend throughout the ending of the interval. A common increase in temperature illustrated by the TI leads to pronounced hydrogen-rich organic matter production and burial where the organic matter flux on average is doubled during the black shale sedimentation compared to prior-OAE conditions. A sudden increase in TOC concentrations coevals with the negative shift in the bulk carbon isotopes and implies a quick and drastic change in the sediment record (Fig. 54). The abrupt increase in OM production can be subjected to an increase in surface water productivity and/or better preservation conditions due to reducing and disoxic conditions. The OAE 1b at the Vocontian Basin is marked by enhanced- and more high-frequency fertility fluctuations in the surface waters, whereas fertility in the OAE 1b from Site 545 was reduced (Herrle 2002). The composition of the OM changes from terrestrial to more marine material as indicated by the Hydrogen Indices (HI; Fig. 38). High HI may contrariwise indicate better preservation conditions as the material could have been often degraded prior to the OAE displaying lower values. Changes in primary productivity or preservation were probably induced by climatic changes.

depth [m]	Lyco*/C31
0,03	0,668
3,63	0,099
4,23	0,711
5,63	0,166
6,93	0,204
7,93	1,291
8,40	1,560
8,44	1,151
8,47	1,757
8,49	2,484
8,90	1,451
9,26	2,586
9,30	1,078
9,38	0,779
9,72	1,087
9,90	0,912
9,95	0,249
9,97	0,906
9,98	3,171
10,00	0,871
10,18	1,253
10,73	0,047
13,13	0,366
13,83	0,143
mean value	1,372

Fig. 65: Excel worksheet of the lycopane*/n-C31 ratio. The marked values belong to the OAE 1b interval.

Specific markers for archael input to sediments, as e.g. lycopane, are abundant and confirm changing water- and environmental conditions. The isoprenoid lycopane is only present during the OAE interval and is seen as a proxy to assess palaeo-oxicity because it is Lycopane is only present during the OAE interval and is seen as a proxy to assess palaeo-oxicity because it is associated to anoxic conditions during sediment deposition (Peters et al. 2005, Wagner et al. 2004, Beckmann et al. 2008). The lycopane*/n-C31 ratio ranges from 0.05 to 3.17 (Fig. 65) approaching highest values during the black shale event (average value=1.37) suggesting oxygen depleted to anoxic conditions (Beckmann et al. 2008). Nevertheless, during the event a few lower values ranging around 1 suggest at least short termed rather suboxic to oxic conditions. Another indicator for methanogenic archaea bacteria are the isoprenoids TMI, PMI and ETMI all present at the Vocontian Basin and other similar sites (see 5.2.3.). Strikingly, the limits of all analysed acyclic isoprenoids exceed the assumed range of the OAE interval (Fig. 66).

	depth [m]	TMI	PMI	ETMI	C23+ETMI
VB80	7,93	1,046	0,984	0,241	
VB83	8,08		0,195		
VB87	8,28		0,084		
VB88	8,33		0,079		
VB89	8,35				
VB90	8,40	0,117	0,152		0,264
VB91	8,44	0,612	0,741		0,943
VB93	8,47	0,638	0,626	0,094	
VB95	8,49	1,679	1,541		1,543
VB101	8,64	0,418	0,303		0,594
VB105	8,72	5,131	4,208		2,389
VB109	8,90	0,337	0,266	0,160	
VB118	9,17	0,525	0,467		0,707
VB122	9,26	0,546	0,412	0,154	
VB124	9,30	1,424	1,345	0,223	
VB128	9,38	1,433	1,568	0,412	
VB132	9,48	0,190	0,787		0,511
VB139	9,72	0,384	0,533		0,654
VB143	9,86	0,199	0,340		0,523
VB145	9,90	0,204	0,293		0,521
VB146	9,95	0,430	0,718		1,022
VB147	9,97	0,863	0,759		0,974
VB148	9,98	0,151	0,186		
VB149	10,00	0,141	0,173		
VB150	10,03	0,267	0,274		0,361
VB153	10,08	0,109	0,128		0,344
VB157	10,18	0,077	0,088		
VB161	10,38	0,006	0,079		
VB163	10,73		0,071		
VB165	10,93		0,046		

Fig. 66: Worksheet displaying the range of the isoprenoids TMI, PMI and ETMI present at the VB. Marked values represent the OAE 1b interval. Note, the isoprenoids exceed the limits of the black shale.

Although their presence is supposed to be restricted to the OAE period, an expansion of the interval duration is questionable due to the record of organic and inorganic parameters. Regarding the concentration trends of e.g. TOC, CaCO3, Si/Al and their flux rates as well as the carbon isotopes, the initiation of OAE 1b is not distinctly visible but clearly the end of OAE sedimentation is. Obviously, methanogenic bacteria populated this palaeo-environment in pre-anoxic conditions but per definition they are restricted to strictly anoxic environments. Thus is seems unlikely that the Vocontian Basin was completely anoxic during the presented interval but that the methanogenic archaea invaded into the Basin when bottom water oxygenation levels were lowered enough during OAE 1b.

Nevertheless, their presence is confirmed as well as the three re-population events of benthic foraminifera (opportunistic species; Friedrich et al. 2005) implying at least short-term periods of enhanced bottom water oxygenation. During these events the TI increase indicates cooling episodes probably due to deep water formation. Oxygen deficient but not anoxic bottomwaters are also indicated by the relatively low abundance of redox-sensitive trace metals. Similar to the MP it can be assumed that improved preservation conditions and enhanced organic matter production fostered the enhanced accumulation of organic matter during OAE 1b (Wagner et al. 2007). Contrary, the most important factor for OAE sedimentation at the Vocontian Basin was probably better preservation conditions, whereas enhanced OM burial was therefore more or less the result of reducing conditions instead of highly increased productivity. This assumption is confirmed by the HI values and the calculated linear sedimentation rates (SR, in cm/ka) which remain fairly constant during the entire sample-record. SR are in general elevated during the entire record (see 5.2.5.) and strikingly values don't increase with commencing OAE 1b but rather decrease during the middle part of the event (Fig. 54).

This short SR collapse is recorded in almost all ratios and certainly in all flux rates from the analysed parameters. Accumulation- and concentration rates from alumo-silicates (Al, Si, K; Fig. 51) and TOC (Fig. 54) show decreasing values whereas coevally the ratio of Si/Al (Fig. 47), carbonate flux rates (Fig. 54) and the ratios of redox-sensitive TM to Al (Fig. 52) increase in concentrations. The drop in sedimentation rates and OM accumulation is almost contemporaneous with one of the re-population- and therefore re-oxygenation events at the Vocontian Basin (see Friedrich et al. 2005). During this interval preservation conditions are lowered due to higher oxygen levels while the detrital input and productivity remain stable or even decrease (see Si/Al ratio and clastic flux rates; Fig. 47 & 51). Similar to pre-OAE conditions no co-variation between organic matter and Zr content is observed during the OAE 1b interval (Fig. 59) supporting a decoupling of organic matter production from wind derived terrigenous matter indicating no direct connection. During black shale sedimentation the concentration of TOC is doubled while Zr values are in the same range as prior to the event. This also might point to better preservation conditions as all parameters reveal nearly "normal" concentrations and ratios but only the TOC is incoherently elevated. A slight negative co-variation confirms the decoupling statement and the low importance of strong wind systems when leading to elevated fertility and/or enhanced preservation conditions. What obviously changed during the event is the enhanced magnitude, displayed by various parameters, pointing to stronger climate contrasts. As mentioned before, the cyclicity remain constant, showing higher contrast with elevated values in both directions, increasing and decreasing concentrations. As consequence, the composition of the sediment does apparently not change. Same trends are discussed before (see 5.3.1.) but the difference during the OAE 1b sedimentation is a stronger affected environmental system.

Affected by intensified climate conditions with extremely humid and extremely arid periods. Climatic changes leading on the one hand to warmer and highly humid conditions with an accelerated hydrological cycle (Herrle 2002) fostering higher organic matter accumulation and/or preservation. During warm/humid intervals the decrease in the clastic flux rates of Si and Zr confirm therefore lower wind velocities. On the other hand during arid and probably more cooling intervals evaporation highly exceeded precipitation at the Vocontian basin and the water mass balance was most likely extremely negative. These assumption are affirmed by the general trends of Si and Al. The ratio confirms that cyclicity remain stable but the amount of Si is compared to Al increased during arid conditions regarding conditions prior to the event (Fig. 67). Nevertheless, looking at the flux rates for these elements, the trend shows decreasing values for both elements and most of the other parameters (Fig. 51). As exceptions are mentioned the organic carbon- and carbonate carbon flux and the CIA (Fig. 51 & 54). The geochemical record presents shifts to humid conditions in nearly the same magnitude as prior to the event, but shifts pointing to arid conditions reveal enhanced magnitude suggesting in general extremely dryer periods during OAE 1b.

Short-termed repopulation events (Friedrich et al. 2005) during the formation of OAE 1b, are supposed to represent re-oxygenation events of bottom waters. Increase of deep water production occurred during cooler intervals when evaporation exceeded precipitation leading to well ventilated oceanic deep water masses (Herrle 2002). Due to high evaporation and small precipitation rates, the hydrological cycle slowed down and sediment transport via fluviatile runoff diminished. The record of the OM production from the Vocontian Basin was mainly controlled by humidity. High rainfall resulted in high fertility and accumulation rates whereas low amounts of rainfall resulted in deep water formation due to high evaporation (Fig. 63). Local wind systems were probably important for ocean circulation patterns and export production but played a minor role for enhanced detrital supply during the event.

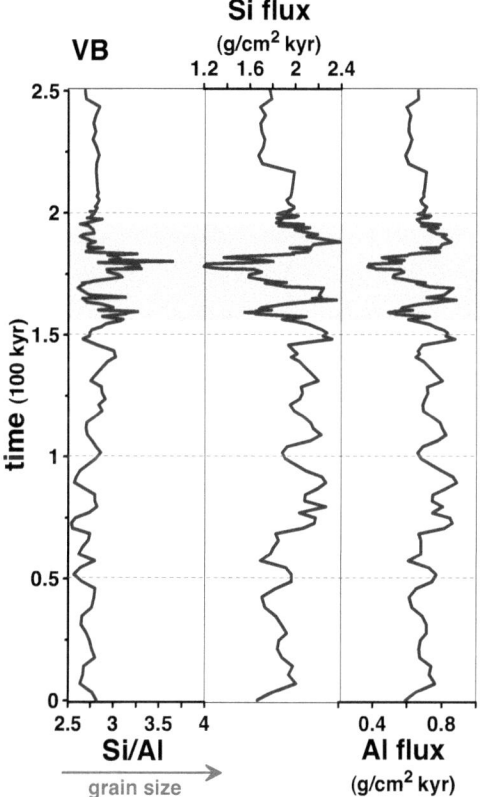

Fig. 67: Si/Al ratio compared to the flux rates of Si and Al shown in the time model. The ratio confirms Si enrichment during arid climate conditions but flux rates display no decisive increase during the black shale event.

Redox-sensitive TM show diverse behaviour (Fig. 52). While the V/Al ratio is enriched during OAE 1b and stepwise decreasing towards the ending, the Ni/Al ratio is highest at the beginning of the event and ranging afterwards at values and fluctuations comparable to pre-OAE values. Zn/Al ratios otherwise even decrease periodically during the event ranging below AVS values. In oxic marine environments Ni and Zn behave as micronutrient. Under reducing conditions (anoxic/euxinic), Ni is incorporated into pyrite or geoporphyrins while Zn may be incorporated as ZnS as a solid solution phase in pyrite or, to a lesser degree, it may form its own sulfides, sphalerite ([Zn,Fe]S; Tribovillard et al. 2006). Vanadium as a conservative TM element implies by its increase the decrease in detrital input while Zn and Ni reflect the geogenic background, contributed by freshwater.

Trends resulting from the A-N-K ternary plot are hardly visible as all samples plot extremely narrow to the A Apex (Fig. 42). During the event samples plot closer to the A Apex as they did prior to it and confirm the CIA, pointing to a higher degree of weathering. During the black shale interval iron concentrations are lowered drastically from values around 3.2 to values around 2.4 (Fig. 41) and coevally the preceding trend is changing: prior to the event iron and sulphur reveal in general the same trends in de- and increasing fluctuations but during the event one parameter increases and the other decreases simultaneously (Fig. 40). The iron content is lowered while the sulphur concentration increases compared to values prior to the event. Increasing sulphur concentrations illustrate once more the better preservation potential attended by the OAE 1b beginning. This also includes reducing iron values due to missing oxidation and resulting in a pausing redox cycle.

The cyclic sedimentation pattern documented prior to the event remains stable without drastic break-down of any parameter. The magnitude of these fluctuations is though enhanced pointing to climate conditions becoming abruptly extreme. In addition, samples display in general a trend to higher aridity during the event. Based on various indications the sedimentation of OAE 1b was probably possible due to better preservation conditions than increased PP. Elevated PP is observed at the VB for the entire record. The sediment composition does obviously not change. A plausible reason for the OAE formation could have been a general (regional) increase in SST with the onset of the black shale leading at this geographical setting to amplified organic matter burial. Because PP was in general high at the VB the surplus production might have fostered better preservation conditions. A proof for higher SSTs at the onset of OAE 1b were brought by the TEX_{86} data from Site 545 and seem to be a logical explanation resuming from the entire data record.

5.3.3. The post OAE stage

At the ending of OAE 1b sedimentation all parameters record the drastic change with an disturbed record. Again, as exception the TOC content drop either immediately at the end of OAE 1b to values comparable to the pre-stage while most of the remaining parameters already decrease in the upper part of the fading OAE sedimentation as e.g. the carbon isotopes. After the event the cyclic fluctuation pattern is disturbed for the first time in the entire record. This is interesting, as cyclicity even remained stable during the event. Nevertheless, the returning to normal conditions after the black shale is obviously accompanied by a drop down in sediment delivery or cyclicity for a comparatively long time period. The disturbance lasted approximately as long as the OAE itself, assuming ~44 kyrs (see 5.2.4. and Fig. 50) for this interval. In the upper part of the record the well known precessional variations reconstituted within similar range as prior to the event. During the entire sediment record all measured parameters range within constant fluctuations changing only in magnitude (Fig. 68).

This might be another evidence for the assumed sedimentation cycle at the VB. A stable eco-system is probably affected by a sudden rise in temperature. Due to its palaeogeographical setting, this increase was coupled to strong climatic contrasts but remaining at stable variations. The returning to normal conditions is contrariwise couppled primary to disturbances. Mainly the inorganic record clearly highlights that the termination of OAE 1b affected the geochemical record stronger than its beginning. Reasons and mechanisms for these special conditions are currently poorly understood and may be solved in continuative working programs. Similar to the MP the time of disturbance is longer than the event. The top 2 metres of the record reveal similar variations as prior to the event suggesting an overall return to pre-OAE conditions. The observed perturbation of the Mediterranean climate system resulted in amplified climate contrasts persisting longer than the isotope excursion similar to Site 545. This may proof the evidence that the Cretaceous land–ocean climate responded with a significant lag in time to atmospheric heating and/or cooling as suggested for the MP.

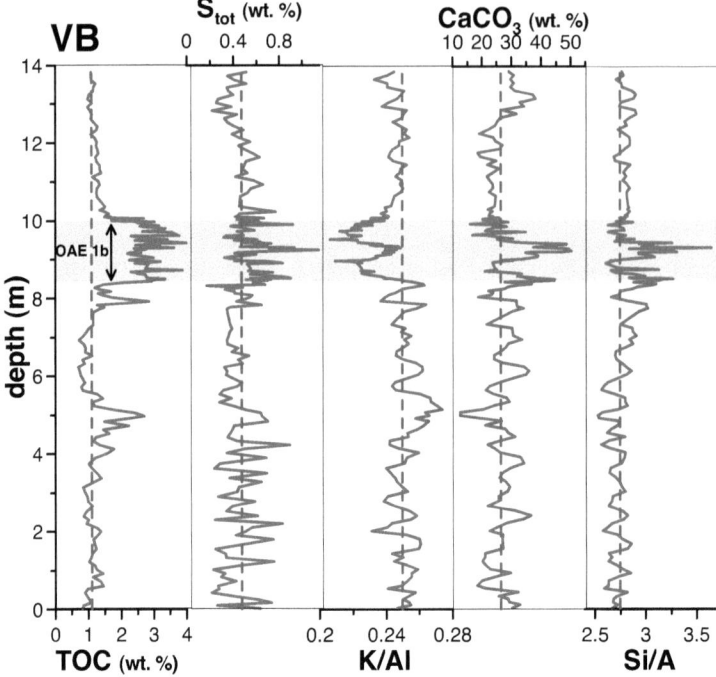

Fig. 68: Selected element ratios and elements displaying constant fluctuations for the entire VB record lying within the red dashed line. Variations are most dominant during the OAE 1b interval but values observed prior to the event are in general obtained again after the event.

The perturbation phase after the event may point to decreasing temperatures attended by decreasing climate contrasts. Based on the TOC record a sudden increase in SST at the beginning of OAE 1b sedimentation is assumed. Likewise is the drop in temperatures at the end of sedimentation. It was speculated that increasing temperatures fostered wind-systems in arid periods and PP as well as higher rainfall in humid periods. When temperatures drop, as sudden as they rose, at the end of OAE 1b this signifies a disturbance of the biosystem obviously because climate contrast obliterate. This may have resulted in a combined climate situation consisting of moist and warm conditions comparable to Site 545 or rather cooler and arid conditions. Which conditions where favoured during the disturbed interval are hard to reconstruct on the basis of the presented data record. But it is obvious that conditions have changed.

In the following Synthesis Chapter predictions about climatic scenarios affecting palaeoenvironments at regional or global scale and mechanisms causing OAEs will be discussed.

6. Synthesis

In the previous chapters of this study the the Mazagan Plateau and the Vocontian Basin, recording the same OAE, were presented and discussed in detail. In the following chapters the focus lies on the linkage of similarities and to figure out differences as well as concerning mechanisms leading to the formation of OAE 1b. Results presented in chapter 6.1. are in general published in Wagner et al. 2007.

6.1. The Trigger of OAE 1b

After introducing chapters presented general possible mechanisms for OAE formations (see e.g. Chapter 1.3.) the specific trigger for OAE 1b is still unclear. To highlight this scenario a coupled atmosphere–land–ocean–sediment model was developed to simulate changes in carbon and nutrient cycling over OAE 1b at the Mazagan Plateau (see Wagner et al. 2007 for more details). The most striking difference to other Cretaceous OAEs is the pronounced negative excursion in $\delta^{13}C_{carb}$ preserved in the Vocontian Basin (Herrle et al. 2004), the Mazagan Plateau (Herrle et al. 2004) and the Northeast Atlantic (Erbacher et al. 2001). Cretaceous OAEs typically show a positive excursion both in $\delta^{13}C_{carb}$ and $\delta^{13}C_{org}$ as a consequence of enhanced global organic carbon burial (e.g., Jenkyns 2003; Leckie et al. 2002).

The parallel negative excursion of carbonate and organic carbon at the base of the OAE 1b is comparable to features reported for the Toarcian (Hesselbo et al. 2000; Jenkyns 2003) and early Aptian OAE 1a (Jenkyns and Wilson 1999; Jahren et al. 2001), and the PETM (Norris and Roehl 1999; Thomas et al. 2002). These features have been attributed to rapid and massive release of isotopically light methane from marine gas hydrates (Jenkyns 2003). Other mechanisms including volcanic outgassing during to the formation of large igneous provinces (Larson & Erba 1999; Leckie et al. 2002) and widespread remineralisation of OC through volcanic intrusions into marine sediments or desiccation of shallow marine areas or continental weathering (Higgins & Schrag 2006), massive biomass burning (Finkelstein et al. 2006) and recycling of respired OC under conditions of a strongly stratifies ocean (Kuespert 1982; Schouten et al. 2000) have also been discussed as drivers of negative carbon isotope excursions and mechanisms of past climate change, albeit not always at the millennial and shorter time scales discussed here.

The modelling setup is forced by methane emission, terrestrial P (Phosphorous) input and burial efficiency, the latter two as a consequence of continental erosion and clastic fluxes to the ocean. As output parameters the model calculates for example the partial pressure of CO_2 in the atmosphere (pCO_2) and corresponding air temperature, export production, oxygen levels and organic carbon burial in sediments, and $\delta^{13}C_{org}$ and $\delta^{13}C_{carb}$ in a given time frame. Two different scenarios were conducted. Scenario 1 simulates emission of methane (Fig. 69), scenario 2 remineralisation of OC (Fig. 70) as trigger for the negative carbon isotopic excursion recorded in the Mazagan record. The results on methane emission achieved the closest match with the geological records.

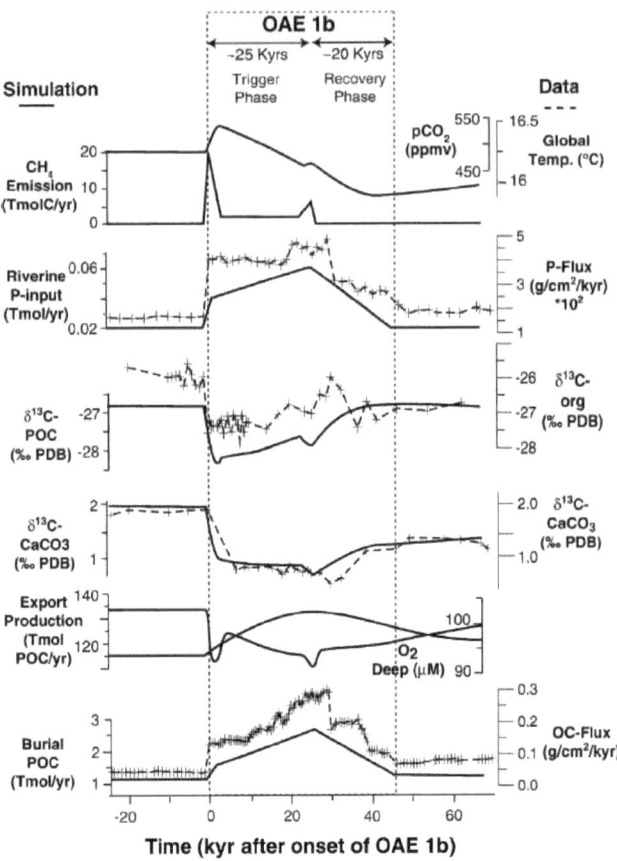

Fig. 69: Results from coupled atmosphere–land–ocean modeling (bold lines with scales on the left, except for uppermost record where scales on the left and right belong to the same modeling result) simulating the obtained geological records of OAE 1b at Mazagan Plateau (stippled lines with scales on the right). Note scales of y-axes are identical for bulk carbon isotopes but differ for the flux records due the different nature and dimension of the two data sets. **Scenario 1**: As primary mechanism the model simulates two moderate pulses of methane emission at the onset and termination of the trigger period with intermitted elevated methane fluxes. Simulated consequences of this trigger phase on climate, continental runoff, marine carbon production and burial, and oxygenation of the deep ocean are displayed showing the complex interwoven relationships between the atmosphere, the land surface and the ocean finally leading to elevated preservation of organic carbon in the sediments (Wagner et al. 2007).

Recycling of respired OC under conditions of a strongly stratified ocean has been shown to cause negative shifts in carbon isotopes accompanied by elevated OC burial (Kuespert 1982). This mechanism is acknowledged but the presented data show no indication of strong stratification or oxygen depletion prior to OAE 1b. The V/Al data from DSDP Site 545 are in the same range before, during and after the event not supporting changes in deep ocean oxygenation throughout the study interval.

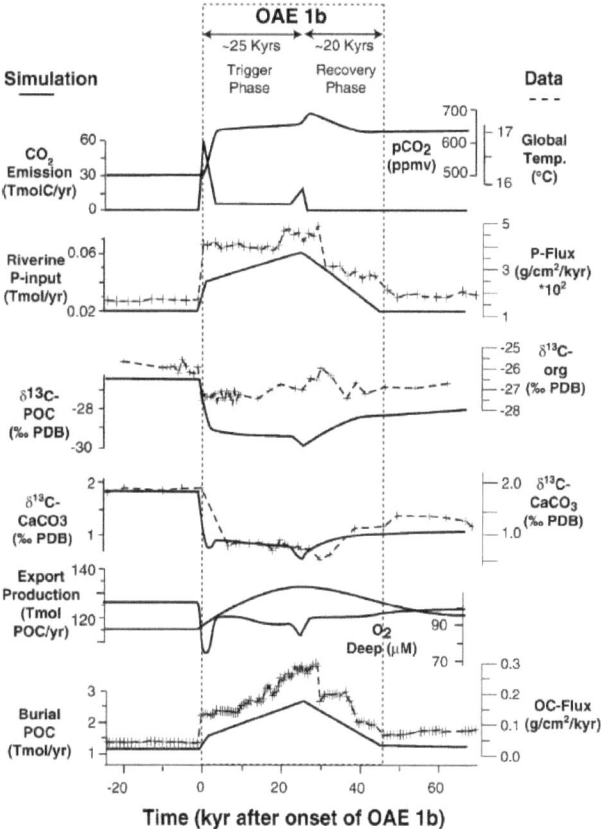

Fig. 70: Same annotation as in Fig. 69. **Scenario 2**: As primary mechanism the model simulates remineralisation of organic matter at the onset and termination of the trigger period. Comparison with the records from the Mazagan site does not show a good match, in particular for bulk carbon isotopes. Simulated pCO_2 and temperature levels remain high throughout and following OAE 1b contradicting climatic evidence from the Vocontian Basin (Herrle, 2003; Herrle et al., 2003). It is therefore suggested that remineralisation of organic matter does not consistently explain the geological records (Wagner et al. 2007).

Following the methane emission scenario 1, a total of 1.15×10^{18} g methane carbon ($\delta^{13}C=-60‰$) is sufficient to generate the recorded $\delta^{13}C_{carb}$ profile. To put this number into perspective, estimates for the gas hydrate inventory at the modern seafloor range in between $1-10 \times 10^{18}$ g methane carbon (Kvenvolden 1998; Milkov 2004). The calculated methane release is thus not unreasonably large compared to modern values but demands that a significant fraction of the early Albian inventory was destabilized during the event. Alternatively, three times the amount of POC ($\delta^{13}C=-25‰$), equivalent to 3.5×10^{18} g C must have been demineralised to generate the observed isotopic excursion. Stratigraphic records from known OAE 1b locations do not sufficiently support this pathway despite evidence for locally enhanced terrigenous matter supply off NW Africa.

Therefore, a methane mechanism is favoured to explain the observed carbon isotopic excursion of OAE 1b. Different from other studies, however, not a single (catastrophic) pulse of methane is assumed or a succession of pulses focused at the onset of the event (e.g. Kemp et al. 2005) followed by a longer phase of recovery. Instead a combination of moderate-scale methane pulses supplemented by continuous methane emission at elevated levels over a longer time interval is invoked. The two pulses, with the initial releasing about twice the volume of methane compared to the terminal one, encompass the total trigger period of OAE 1b, estimated to be about 25 kyr.

Following this scenario, an initial pulse of methane may have caused the observed rapid negative shift in $\delta^{13}C$ at the onset of OAE 1b. The initial methane peak lasted about 4 kyr and released 0.52×10^{18} g methane carbon into the ocean as deduced from the model simulation. A maximum methane emission of 20 Tmol CH_4/yr during this period is high in comparison to modern methane flux at the seafloor, estimated to be 2–3 Tmol/yr (Kvenvolden and Lorenson 2001) while mud volcanoes on land emit about 0.4–0.6 Tmol/yr (Etiope and Milkov 2005). Accordingly, methane emissions were enhanced by about one order of magnitude during the initial period of OAE 1b. After the initial set-off of the event, methane continues to emit at a level of 2 Tmol/yr to maintain the negative ^{13}C values recorded in marine carbonates and sedimentary organic matter. This methane flux is very moderate, comparable to the modern seafloor flux, and as one possibility may represent progressive melting of a hydrate-rich continental margin until the field is depleted. Once methane emissions, and thus the trigger, terminate following a final pulse of methane the climate system instantaneously starts to recover and approaches almost pre-event conditions after another 15–20 kyrs. The total amplitude of global climate variability during

OAE 1b is small, less than 100 ppmv CO_2 or about 0.8°C global average air temperature, its duration between 35–40 kyr. Still in a warmer world this variability is sufficient to perturb the Earth system over orbital time scales and, as recorded at Mazagan Plateau, stimulate a series of strong regional feedbacks. This interpretation is also supported by the divergence between constant $\delta^{13}C_{carb}$ and increasing $\delta^{13}C_{org}$ values during the trigger phase of OAE 1b observed both in the data and the model results. $\delta^{13}C_{org}$ values depend on the concentration of dissolved CO_2 in surface water (Popp et al. 1989) and therefore mirror the evolution of atmospheric pCO_2. They start to recover immediately after the initial negative excursion caused by the proposed methane pulse since CO_2 produced by methane oxidation is rapidly consumed by a combination of enhanced marine export production, silicate weathering and reduced carbonate accumulation.

The results from the three OAE 1b locations, the Vocontian Basin and the Mazagan Plareau presented in this study and Site 1049 off Florida (Erbacher et al. 2001), suggest that subtropical areas experienced a synchronous rapid change in climate and ocean structure, supporting a global nature of the proposed Greenhouse gas (GHG) emission event. Perturbation of marine hydrates is favoured in this study to be the main trigger of OAE 1b, though it is acknowledged that other mechanisms can cause comparable geochemical signatures in the geological record. Despite uncertainties in providing a final proof of the trigger mechanism, perturbation of marine hydrates is a well recognized geo-hazard and the case study presented in Wagner et al. 2007 may emphasizes their exposure. Once set-off, disintegration of marine hydrates can not be stopped until the reservoir is depleted.

6.2. Comparison of both sites

6.2.1. Differences and similarities

As already discussed in the preceding chapters the list of major differences between both sites is long. Avoiding recurrences, only main climatic varieties referring to regional distinctions are mentioned (Fig. 13). Site 545 at the Mazagan Plateau is situated open marine at the Cretaceous Atlantic. Its geographical position at the southern tropical belt is coppled to a negative water mass balance where evaporation exceeds in general precipitation. Like nowadays, PP was mainly driven by coastal upwelling which was due to strong western trade winds very high. Therefore, the aeolian delivered fraction, composed of Si, Zr and Ti, and continental weathering was also high. On the continental slope at palaeo-water depths around 2000 m, Site 545 reflects a undisturbed deep-water environment where sedimentation reflects nearly normal marine conditions for the opening Atlantic and the terrigenous input from the African Hinterland delivered mainly by westwards directed wind systems.

The Vocontian Basin was located at the northern tropical belt, influenced by the Cretaceous Tethyan Ocean. For this area higher precipitation than evaporation rates are assumed. Local wind systems are strong during arid periods but without affecting coastal upwelling. Although deep water formation took place in the deeper part of the Tethyan, deep ocean currents were probably less important than at the Mazagan Plateau. At the Vocontian Basin wind driven surface currents may have played a dominant role. The interactions within the Mediterranean area are complex during the middle Cretaceous (Fig. 57). The sample record represents in general a pelagic basin of several hundred meters surrounded by drowned carbonate platforms reflecting high tectonic activity as e.g. subsidence and uplift phases and relations to a pull-apart basin. Contrary to Site 545 the Vocontian Basin section was not primary influenced by the hinterland and its terrigenous fraction. The sample record reflects the deeper part of a pelagic basin, though this was anyway shallow compared to Site 545, and no exposed land mass was probably situated nearby (see Fig. 35).

The onset of OAE 1b is documented by the negative excursion in $\delta^{13}C_{carb}$ and $\delta^{13}C_{org}$. Here, the recorded isotope signature of this event is different at both sites. At Site 545 a sudden drop to negative values remain constant lasting around 45 kyrs over the entire interval. This negative shift is so drastic that the possibility of sedimentary gaps became obvious. But the remaining geochemical record as well as sedimentological analysis indicate a consistently sedimentation without gaps. The negative excursion remaining constant for the entire OAE interval is, as discussed, interpreted to result from the continuous emission of marine hydrates at elevated levels over a longer time interval. A initial and a terminal pulse are thought to be responsible for a constant methane emission explaining the negative isotope excursion. Comparing these results with the negative excursion at the Vocontian Basin seems to be difficult at the first look because of the two-folded peak. But at the second look and combining the model results from the Mazagan Plateau (Wagner et al. 2007) with the Vocontian Basin, strikingly the second peak in the isotopes goes synchronous with the modeled terminal methane pulse (Fig. 71). This observation supports the presented methane-model for Site 545. Obviously, the magnitude of the two carbon isotope shifts is pronounced and the general trend is, compared to Site 545, strongly fluctuating at the Vocontian Basin but in the same order as the entire record does.

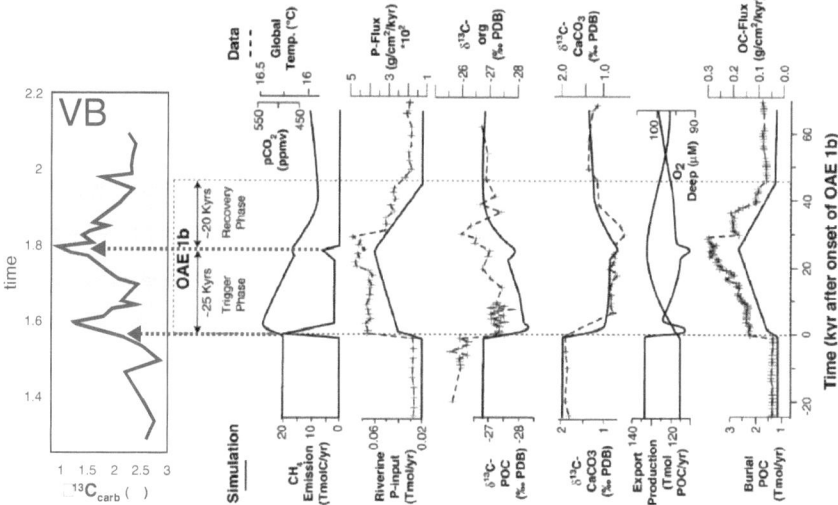

Fig. 71: The bulk carbon isotope record from the VB in comparison to the methane-model (scenario 1; Fig. 69) calculated for Site 545 (the Mazagan Plateau). The two shifts in the isotope record are in good correlation to the estimated methane pulses (after Wagner et al. 2007).

The isotope signature at the Vocontian Basin records obviously the same event under different conditions. Reviewing the geochemical data set of both sites this difference is already visible in the interval prior to the OAE and remains stable during the post-OAE stage. While the black shale deposition at the Mazagan Plateau is described as a productivity event, evidence at the Vocontian basin support sedimentation due to enhanced preservation conditions. This suggestion is also supported by the stepwise TOC increase at the Mazagan Plateau and the sudden increase at the Vocontian Basin. The general sedimentation pattern at Site 545 is furthermore different during the event and shortly after it than prior to OAE 1b, supporting strongly changing climate conditions. At the Vocontian Basin, on contrary, the trend reveals equal but with enhanced magnitude where only the carbon isotopes and the TOC content document the black shale event with an significant changing record. The discussed changing climate conditions at Site 545 affect also the mediterranean Area but without drastic changes. Here, climate contrasts are becoming stronger for the time of enhanced carbon burial. The ending of OAE 1b is therefore clearly visible in all parameters. As proposed for Site 545, the fading OAE sedimentation is accompanied with a beginning cooling interval. All inorganic geochemical parameters record this condition as disturbance for both sites.

Resuming, both sites document the same event but react different due to regional distinctions before, after and during OAE 1b. A comparable period is when shortly after the OAE sedimentation both records reveal a similar time lack of environmental perturbation. At the Mazagan Plateau the perturbation commences already with the beginning OAE 1b. The sudden negative shift in the carbon isotopes, documented already by various researchers, is together with the instantaneous increase of the TOC content the most prominent and relevant feature both sites have in common. Although the isotope record during the negative excursion differs apparently, the main trends are the same. The first negative shift is abrupt and remains constant at Site 545 while it decreases directly at the Vocontian Basin to values comparable prior to the event. The second pulse and most negative values at Site 545 is accompanied by a slight shift to lighter carbon isotopes. At the Vocontian Basin this second pulse, documenting most negative values as well, has the same magnitude as the first one. It does not remain constant but decreases immediately towards the ending of OAE 1b sedimentation. Obviously driven by different orbital frequency bands, the record of the VB is changing with a high frequency (precession) and the record of the MP reacts to a superordinated orbital band (eccentricity). Changing solar radiation, earth rotation and a shifting ITCZ had determined more or less drastic consequences on the respective environmental system.

6.2.2. Regional distinctions and global climate trends

The VB was highly influenced by precessional orbital forcing showing high variability in shorter time periods than the Mazagan Plateau which was mainly influenced by the eccentricity frequency band also driving the general temperature trends. The main climate varieties are the differences in the water-mass balance. Average solar radiation was higher at the MP reflecting high aridity. Here, obviously the hydrological cycle was controlled by strong wind systems over the subtropical hemisphere, nowadays comparable with the northeastern tradewinds, leading to high aridity which exceeded precipitation. Nevertheless, these tradewinds fanned at the MP coastal ocean currents leading to coastal upwelling and high PP. The onset of OAE 1b was accompanied by increasing temperatures leading in an already extreme warm global situation to drastic climatic changes into greenhouse conditions. For the Mazagan Plateau this temperature increase followed by the black shale sedimentation also led to a positive water mass balance probably due to a shifting ICTZ (see 4.3.2.).

Ufnar et al. (2004) showed modeled Albian precipitation and evaporation rates for the North American Cretaceous Western Interior Basin (WIB). The authors used the precipitation (P) and evaporation (E) fluxes to calculate the net moisture flux (P-E balance) value for each incremental step in palaeolatitude. The calculated P-E balance and precipitation rate were then used to calculate the net moisture fluxes in millimeters per year. Net moisture fluxes (in mm/yr) were then used to calculate the latent heat values. Their results showed, that a comparison of modeled Albian and modern precipitation minus evaporation values suggest amplification of the Albian moisture deficit in the tropics and moisture surplus in the mid to high latitudes. The presented results of this study support this prediction. As Ufnar et al. (2004) calculated the net moisture flux for the Western Interior Basin in North America, a direct comparison is difficult and only parallels can be considered.

Therefore, the net flux rates for the Turonian (93 Ma) were calculated on a selected section (0-80°N 26-34°W) comprising the Mediterranean Area as well as the Mazagan Plateau (Data kindly provided by S. Floegel, pers. communication; Fig. 72). Sedimentologic interpretation of the results of numerical simulations using the GENESIS atmospheric general circulation model were used to determine what effect orbital cycles might have had on sedimentation (see Floegel et al. 2005 for more details on the model). Some specific features of GENESIS are the use of an Atmospheric General Circulation Model (AGCM) as its core component which is coupled to a 50-m mixed slab ocean model and multilayer models of soil, snow and sea ice. Resolution of the AGCM is 3.75°x3.75° and that of the Land Surface Transfer Scheme (LSX) is 2°x2° (Floegel et al. 2005).

Discussing the results it should be kept in mind that the model represents a time frame ten million years later than sediments from the MP and the VB. Also the palaeogeography changed due to the beginning closure of the SFB with the beginning Turonian (Arnaud & Lemoine 1993). This time period is also characterised by the late Cretaceous folding and the Vocontian trough was nearly entirely sedimented. Due to rifting in the North Atlantic and in the southern Tethys, Europe was already moved by about 3-5° of latitude northwards in the middle to late Albian (Voigt et al. 1999). Nevertheless, a high range of similarities are expected due to a stable northern reef line at 30°N from the late Early to the latest Late Cretaceous, suspecting that the mean annual position of the subtropical highs was consistently located about 30°N in the Mediterranean region (Voigt et a. 1999).

The annual mean of the Turonian data set reveals an arid zone extending from ~10°-30°N palaeolatitudes (Fig. 73). After Ufnar et al. (2004) highest evaporation rate and aridity is for the WIB predicted between 10° and 25°N palaeolatitudes comprising the location of Site 545. Voigt et al. (1999) also assumed that the greatest deficit of P-E was situated above the tropical-subtropical areas of the mediterranean Tethys south of the subtropical highs between 10 and 30°N palaeolatitude. The Vocontian Basin was situated between 25° and 30°N and the map shows for this area decreasing contrast and smoothing transition from negative to positive net moisture fluxes, which is not surprising assuming a transition from subtropical to midlatitude humid climate north of 30°N (Voigt et al. 1999).

93m6orba

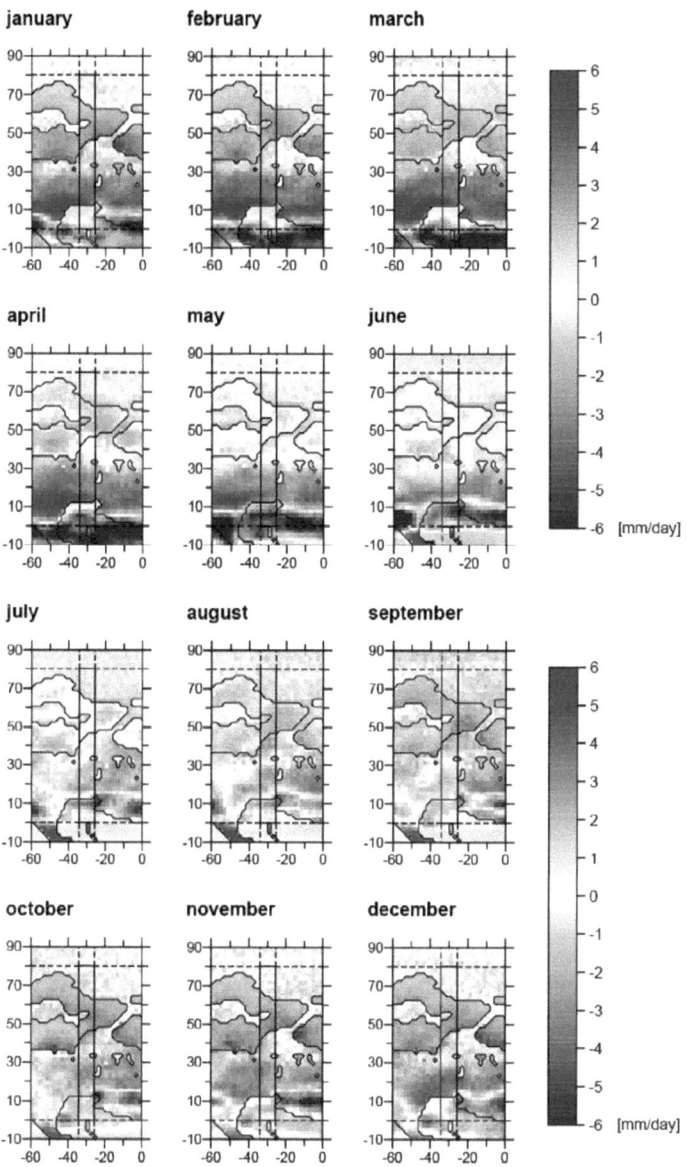

Fig. 72: Net flux rates for the Turonian (93 Ma), calculated on a selected section (0-80°N 0-60°W) comprising the Mediterranean Area as well as the Mazagan Plateau. Numerical simulations were

conducted using the GENESIS atmospheric general circulation model. Red shows settings with negative water mass balance and blue reveals positive values.

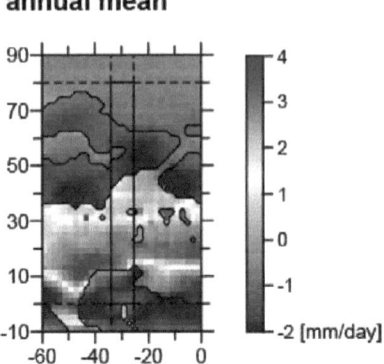

Fig. 73: The annual mean of the Turonian data set. An arid zone is situated in the range from ~10°-30°N.

Based on a plausible xy-plot showing the palaeolatitude against precipitation and evaporation rates (Figure 2, in Ufnar et al. 2004) the same was conducted with the here presented results showing general trends of aridity versus humidity (Fig. 74). The graph shows the precipitation curve as blue line and as red line the evaporation curve (millimeters per year) starting at 0° latitude and ending at 80° N. Between 0°-10°N a positive water-mass balance is expected probably due to the influence of the ITCZ. Slightly above 10°N a negative net moisture flux commences where the tropical to subtropical areas regional begin. The Mazagan Plateau is situated where evaporation exceeds precipitation by the maximum factor (Fig. 74). Influenced strongly by tradewinds it represents extreme dry conditions, whereas this leads to pronounced coastal upwelling and high PP.

At the Vocontian Basin both lines are more approximate reducing the strong flux contrast to a minimum. The graph demonstrates evaporation rates exceed precipitation slightly until 40°N but the divergence is suddenly decreasing once passing the 30°N palaeolatitude. After Hay (2008) the Vocontian Basin was situated in the subtropical highs representing high insolation and surface warming and excluding clouding, condensation and strong wind forcing supporting the presented data set. Seasonal changing climate variations during the Mid-Cretaceous can be expected at the Vocontian Basin due to its exposed position located at the subtropical highs and also as a bordering area at the transition from arid to humid conditions and lagoonal versus estuarine circulation patterns (Voigt et al. 1999).

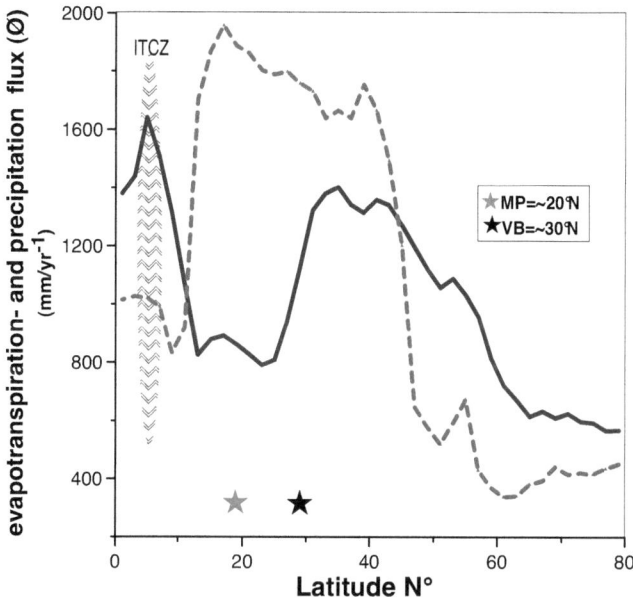

Fig. 74: Modeled precipitation (solid line) and evaporation rates (dashed line) for the Turonian stage. Asterisks mark the palaeogeographical setting of the VB and the MP. The ITCZ was supposed to be situated around ~5°N.

With the beginning of OAE 1b sedimentation the global warming due to supposed methane emissions changed climate conditions at both sites. The general upwelling regime at the Mazagan Plateau broke down due to alleviative coastal winds as a result of increasing humidity. The temperature increase and the initiation of OAE influenced environmental conditions at the Vocontian Basin to a lesser degree but reveals obviously enhanced climate contrasts and a general trend towards higher aridity. The different climate conditions were discussed already in detail for each site but for the first time these conditions can be joined together in an supraregional context. The result of high insolation and the greenhouse warming might be a ITCZ shifting slightly more during OAE 1b sedimentation as during the remaining record. Regarding the P/E graph, it can be assumed that the ITCZ was situated, similar to modern climate conditions, around the equator at 5°N palaeolatitude. The Mazagan Plateau is becoming wet by a ITCZ shift of 15°, which is comparable to modern conditions in the Mediterranean area during summer. It must be pointed out that at least nowadays the position of the ITCZ is strongly influenced by the the different heating of ocean- and land areas. This results in minor position changes for the course of the year of only a few grade over the Pacific and Atlantic oceans. Over huge landmasses, e.g. the Indian continent, the shift is very distinct with around 40°N.

Based on the assumption that during the Cretaceous the position of the subtropical highs was consistent located about 30°N in the Mediterranean region, strong variations of the ITCZ can be excluded. A ITCZ shift of 15° to the North leading to total different environmental conditions was therefore an exemption restricted to the duration of OAE 1b and the effect of a sudden increase of global temperatures. While Site 545 than reveals a positive fresh water balance, the Vocontian Basin nearly reflects the same arid climate conditions as the Mazagan Plateau prior to the ITCZ shift. This shift explains the beginning humid climate conditions coupled to alleviative tradewinds at the Mazagan Plateau and also the enhanced climate contrasts at the Vocontian Basin with a general trend towards higher aridity. At the end of OAE 1b sedimentation and the global warmth period, the system re-establishes normal conditions after a short phase of "rebooting".

6.3. OAE 1b: Regional or a global event

After presenting and discussing the results of this study the question about the expansion of OAE 1b as a global or regional event is still unsolved. Climatic different regions were therefore analysed to see how they react under extreme conditions and extensively discussed. The main difference between OAE 1b and the "big two" events (OAE 1a and OAE 2) are as follows: OAE 1b covers a shorter time period, the OC values are elevated but not as pronounced, same applies for the shift in the carbon isotopes and finally, OC-rich layers from OAE 1a and 2 can be correlated across large distances confirming their world wide distribution.

Nevertheless, suggesting methane emissions as the trigger mechanism for the deposition of OAE 1b this scenario is definitely a global event affecting the atmosphere and changing SST by a global rise in temperature. For sure, OAE 1b was not as pronounced as the "big two" events but the rather constant methane emission from marine hydrates over a probably shorter time period is documented in various deposits comprising the Mediterranean Area and within the opening Atlantic area (see also Chapter 1.4.). Because of the different climate belts and sedimentation cycles the thickness and OC-values of the black shales vary strongly. Due to the tectonic rifting of the Atlantic ocean the marine methane field was probably located within this area assuming also a higher impact for this region, comprising the Mazagan Plateau. Hence, OAE 1b is not documented in the Pacific or Indian Ocean. Still, for the Cretaceous most of the global features occurred when rifting started in the Jurassic-Cretaceous period and the separation of the continental plates lead to the opening of the Atlantic ocean. The lower amount of organic carbon maybe the reason why preservation of the OAE 1b was not possible everywhere because of different redox and oxygenation parameters. Bralower et al. (1999) complained that most of the current understanding of the mid-Cretaceous is based on investigations of pelagic sections from southern Europe and deep-sea drilling sites. They focused on hemipelagic sections deposited along the continental margin of northeastern Mexico. Results were correlated to the OAE 1a, OAE 1b and an event in the late Aptian *Globigerinelloides algerianus* Zone. The authors concluded that OAE 1a and OAE 1b were global events due to the similar C-isotope curves in Mexico and other areas and the response of carbon isotopes to the OAEs.

Results from this study support their thesis, suggesting a constant relationship with an external forcing mechanism. Although Bralower et al. (1999) favoured volcanism, the here presented coupled atmosphere–land–ocean–sediment model (Wagner et al. 2007) rather favours the perturbation of marine hydrates as the most possible trigger. OAE 1b extended from the eastern part of the mediterranean/Tethys area (Ionian Zone, NW-Greece) to the continental margin of northeastern Mexico comprising a large area and numerous locations of black shale deposits. OAE 1b was therefore, as shown, not restricted to specific regional settings, climatic conditions or environmental factors. The organic matter is for example not restricted to a certain bacterial source (which were common in non-upwelling settings, like the Vocontian Basin and the Ionian Basin) nor to high PP due to coastal upwelling. For sure, the main similarity is the restriction of OAE 1b sediments along the Atlantic Ocean or the shallow part of the Tethyan area probably due to simplified ocean currents and limited oxygenated deep water masses.

If OAE 1b was a global, a regional or rather a supra-regional event is clearly a question of definition. But the numerous studies, the here presented included, showed that OAE 1b is very diverse in its formation depending on regional variations but showing lastly for every location the same characteristic features of this black shale event. Even the $\delta^{13}C_{org}$ isotope signature for the OAE 1b from the Santa Rosa Canyon section in Mexico (Bralower et al. 1999) is very similar to the here presented sites. It records a two folded isotopic shift with an amplified second pulse comparable to the Vocontian Basin record and supporting the assumed methane scenario. Resuming, the OAE 1b is recorded from several distinctive settings reflecting specific climatic and environmental conditions still identifiable due to characteristic C-isotope curves and can therefore be stated as a global event.

7. References

Algeo T.J. & Maynard J.B. (2004) Trace-element behaviour and redox facies in core shales of Upper Pennsylvanian Kansas-type cyclothems. Chemical Geology 289, 218–318.

Arnaud H. & Lemoine M. (1993) Structure and Mesozoic-Cenozoic evolution of the South-East France Basin (SFB). Geol. Alpine, Série Spéciale 'Colloques Excursions', 3, 3–58.

Arthur M.A., Jenkyns H.C., Brumsack H.-J. & Schlanger S.O. (1990) Stratigraphy, Geochemistry, and Paleoceanography of Organic Carbon-Rich Cretaceous Sequences. In: Ginsburg R.N. & Beaudoin B. (eds.) Cretaceous Resources, Events and Rhytmus: Dordrecht, NATO ASI Series, Series C. 304, 75-119.

Arthur M.A. & Sagemann B.B. (1994) Marine Black Shales: A Review of Depositional Mechanisms and Environments of Ancient Deposits. Annual Rev. Earth & Planetary Science 22, 499-552.

Arthur M.A. (2000) Volcanic contributions to the carbon and sulphur geochemical cycles and global change. In: Sigurdsson H. (ed.) Encyclopedia of Volcanoes. Elsevier Science (USA), 1045-1056.

Atrops F. (1984).- 2.7. Chaînes subalpines (Jurassique supérieur : Malm). In: Debrand-Passard D.S. (ed.) Synthèse géologique du Sud-Est de la France. Mémoire du Bureau des Recherches Géologiques et Minières, Orléans 1 (125) (Stratigraphie et paléogéographie), 255-257.

Barker C.E., Pawlewicz M.J. & Cobabe E.A. (2001) Deposition of sedimentary organic matter in black shale facies indicated by the geochemistry and petrography of high resolution samples, Blake Nose, western North Atlantic. In: Kroon D., Norris R.D. & Klaus A. (eds.) Wester North Atlantic Paleogene and Cretaceous Palaeoceanography. Special Publications 183. Geological Society, London, 49–72.

Barron E.J. (1992) Lessons from past climates. Nature 360, 533.

Beckmann B., Flögel S., Hofmann P., Schulz M. & Wagner T. (2005) Orbital forcing of Cretaceous river discharge in tropical Africa and ocean response. Nature 437, 241-244.

Beckmann B. (2005) Development and controls of the final Cretaceous black-shale event, Coniacian to lower Campanian (OAE 3): A reference section from the tropical Atlantic at Milankovitch time-scales. Dissertation, Universität Bremen, 99 pp.

Beckmann B., Hofmann P., März C., Schouten S., Sinninghe Damsté J.S. & Wagner T. (2008) Coniacian–Santonian deep ocean anoxia/euxinia inferred from molecular and inorganic markers: Results from the Demerara Rise (ODP Leg 207). Organic Geochemistry 39, 1092-1096.

Bice K.L. & Norris R.D. (2002) Possible atmospheric CO_2 extremes of the Middle Cretaceous (late Albian – Turonian). Paleoceanography 17 (4), 22-1 – 22-17.

Bice K.L., Bralower T.J., Duncan R.A., Huber B.T., Leckie R.M. & Sageman B.B (2002) Cretaceous climate-ocean dynamics : Future directions for IODP. A JOI/USSSP and NSF sponsored workshop. The Nature Place, Florissant, Colorado. Available under: http://www.whoi.edu/ccod/CCOD_report.html#oae_captions

Bice K.L., Huber B.T. & Norris R.D. (2003) Extreme polar warmth during the Cretaceous greenhouse? Paradox of the late Turonian $\delta^{18}O$ record at Deep Sea Drilling Project Site 511. Paleoceanography 18 (2), 9-1 – 9-11.

Bice K.L., Birgel D., Meyers P.A., Dahl K.A., Hinrichs K.-U. & Norris R.D. (2006) A multiple proxy and model study of Cretaceous upper ocean temperatures and atmospheric CO_2 concentrations. Paleoceanography 21 (2), PA2002.

Bigg G.R. (2003) The oceans and Climate. Cambridge University Press, 273 pp., United Kingdom.

Bornemann A., Norris R.D., Friedrich O., Beckmann B., Schouten S., Sinninghe Damsté J.S., Vogel J., Hofmann P. & Wagner T. (2008) Isotopic evidence for a glaciation event during the early Late Cretaceous super-greenhouse episode. Science 319, 189–192.

Bralower T.J. & Thierstein H.R. (1984) Low productivity and slow deep-water circulation in mid-Cretaceous oceans. Geology 12, 614-618.

Bralower T.J., Slitter W.V., Arthur M.A., Leckie R.M., Allard D.J. & Schlanger S.O. (1993) Dysoxic/anoxic episodes in the Aptian-Albian. In: Pringle M.S., Sager W.W. Sliter W.V. & Stein S. (eds). The Mesozoic Pacific: Geology, Tectonics, and Volcanism: American Geophysical Union Monograph 77, 5-37.

Bralower T.J., Cobabe E., Clement B., Sliter W.V., Osburn C.L. & Longoria J. (1999) The record of global change in mid-Cretaceous (Barremian-Albian) sections from the Sierra Madre, Northeastern Mexico. Journal of Foraminiferal Research 29 (4), 418-437.

Bray E. E. & Evans E. D. (1961) Distribution of n-paraffins as a clue to recognition of source beds. Geochim. Cosmochim. Acta 22, 2-15.

REFERENCES

Bréhéret J.G. (1985) Indices d'un énénement anoxique étendu à la Téthys alpine, à l'Albien inférieur (énénement Paquier). Comptes Rendus de l'Academie des Sciences 300 (8), 355-358.

Bréhéret J.G. (1988) Épisodes de sédimentation riche en matière organique dens les marnes bleues d'age aptien et albien de la partie pélagique du bassin vocontien. Bulletin de la Societé Géologique de France 4, 349-356.

Bréhéret J.G. (1994a) The mid-Cretaceous organic-rich sediments from the Vocontian zone of the french South-East Basin. In: Mascle A. (eds.) Hydrocarbon and petroleum geology of France 4, 295-320. The European Association of Petroleum Geoscientists Special Publications, Springer-Verlag, Berlin/Heidelberg.

Brumsack H.-J. (2006) The trace metal content of recent organic carbon-rich sediments: Implications for Cretaceous black shale formation. Palaeogeography, Palaeoclimatology, Palaeoecology 232, 344-361.

Canfield D. (1989) Sulfate reduction and oxic respiration in marine sediments: implications for organic carbon preservation in euxinic environments. Deep Sea Research 6, 121–138.

Chamley H. & Debrabant, P. (1984) Mineralogical and geochemical investigation of sediments on Mazagan Plateau northwestern African margin (Leg 79, Deep Sea Drilling Project). In: Hinz, K., Winterer, E.L., et al. (Eds.), Initial Reports of the Deep Sea Drilling Project, vol. 79. U.S. Govt. Printing Office, Washington, pp. 497–506.

Chamley H. (1989) Clay Sedimentology. Springer-Verlag, Berlin.

Clarke L.J. & Jenkyns H.C. (1999) New oxygen isotope evidence for long-term Cretaceous climatic change in the Southern Hemisphere. Geology 27 (8), 699-702.

Cotillon P. & Rio M. (1984) Cyclicité comparée du Crétacé inférieur pélagique dans les chaînes subalpines méridionales (France SE), l'Atlantique central (Site 534 DSDP) et le Golfe du Mexique (Sites 535 et 540 DSDP). Implications paléoclimatiques et application aux corrélations stratigraphiques transtéthysiennes. Bulletin de la Société Géologique de France, 26, 47-61.

Crowley T.J. & North G.R. (1991) Paleoclimatology. Oxford Monographs on Geology and Geophysics 18, 349 pp., United states of America.

Curnelle R. & Dubois P. (1986) Evolution mésozoïque des grands bassins sédimentaires français (bassins de Paris, d'Aquitaine et du Sud-Est).- Bulletin de Société Géologique de France, 8, 529-546.

Dean W.E. & Arthur M.A. (1989) Iron-sulfur-carbon relationships in organic-carbon-rich sequences I: Cretaceous Western Interior Seaway. American Journal of Science 289, 708-743.

De Leeuw J.W. & Sinninghe Damsté J.S. (1990) Organic sulfur compounds and other biomarkers as indicators of paleosalinity. In: Orr W.L. & White C.M. (eds) Geochemistry of Sulfur of Fossil Fuels. American Chemical Society, Washington, DC, 417-443.

Deroo G., Herbin J.P. & Roucaché J. (1984) Organic geochemistry of Cenozoic and Mesozoic sediments from Deep Sea Drilling Project Sites 544 to 547, Leg 79, Eastern North Atlantic. In: Hinz K. & Winterer E.L. (Eds.), Initial Reports of the Deep Sea Drilling Projects, Vol. 79, 721-742. U.S. Government Printing Office, Washington D.C.

Emeis K.-C. & Morse J.W. (1993) Zur Systematik der Kohlenstoff-Schwefel-Eisen-Verhältnisse in Auftriebssedimenten. Geologische Rundschau 82, 604-618.

Erba E. (1994) Nannofossils and superplumes: the early Aptian "nannoconids crisis." Paleoceanography 9, 483–501.

Erba, E. 2004. Calcareous nannofossils and Mesozoic oceanic anoxic events. Marine Micropaleontology 52, 85-106.

Erba E. & Tremolada F. (2004) Nannofossil carbonate fluxes during the Early Cretaceous: Phytoplankton response to nutrification episodes, atmospheric CO_2, and anoxia. Paleoceanography 19, 10.1029/2003PA000884

Erbacher J., Thurow J. & Littke R. (1996b) Evolution patterns of radiolaria and organic matter variations: A new approach to identify sea-level changes in mid-Cretaceous pelagic environments. Geology 24 (6), 499-502.

Erbacher J. & Thurow J. (1997) Influence of oceanic anoxic events on the evolution of mid-Cretaceous radiolaria in the North Atlantic and western Tethys. Marine Micropaleontology 30, 139-158.

Erbacher J., Gerth W., Schmiedl G. & Hemleben C. (1998) Benthic foraminiferal assemblages of late Aptian-early Albian black shale intervals in the Vocontian Basin, SE France. Cretaceous Research 19, 805-826.

Erbacher J., Hemleben C., Huber B. T. & Markey M. (1999) Correlating environmental changes during early Albian oceanic anoxic event 1B using benthic foraminiferal paleo-ecology. Marine Micropaleontology 38, 7-28.

Erbacher J., Huber B.T., Norris R.D. & Markey M. (2001) Increased thermohaline stratification as a possible cause for an oceanic anoxic event in the Cretaceous Period. Nature 409, 325-327.

Espitalié J., Makadi K.S. & Trichet J. (1984) Role of the mineral matrix during kerogen pyrolysis. Organic Geochemistry 6, 365-382.

Etiope G. & Milkov A.M. (2005) A new estimate of global methane flux from onshore and shallow submarine mud volcanoes to the atmosphere. Environ. Geol. 46, 997–1002.

Ferry S. & Rubino J. L. (1989) Mescoic Eustacy Record on the Western Thethyan Margins –Guidbook of the post-meeting fieldtrip in the Vocontian Trough. Publ. Assoc. Sédim. Fr., 12.

Ferry S. (1990) Mesozoic eustacy record on western Tethyan margins. Post-field comments. Post-meeting field-trip in the Vocontian Trough. Publication de l'Association des Sédimentologistes français 12, 121-140.

Finkelstein D.B., Pratt L.M. & Brassell S.C. (2006) Can biomass burning produce a globally significant carbon-isotope excursion in the sedimentary record? Earth Planet. Sci. Lett. 250 (3–4), 501–510.

Floegel S., Hay W.W., DeConto R.M. & Balukhovsky A.N. (2005) Formation of sedimentary bedding couplets in the Western Interior Seaway of North America— implications from Climate System Modeling. Palaeogeography, Palaeoclimatology, Palaeoecology 218, 125–143.

Föllmi K.B., Weissert H., Bisping M., and H. Funk (1994) Phosphogenesis, carbon-isotope stratigraphy, and carbonate platform evolution along the Lower Cretaceous northern Tethyan margin. Amer. Assoc. Petrol. Geol. Bull. 106, 729-746.

Föllmi K.B., Godet A., Bodin S. & Linder P. (2006) Interactions between environmental change and shallow water carbonate buildup along the northern Tethyan margin and their impact on the Early Cretaceous carbon isotope record. Paleoceanography 21, 1-16.

Forster A., Schouten S., Moriya K., Wilson P.A. & Sinninghe Damste J.S. (2007) Tropical warming and intermittent cooling during the Cenomanian/Turonian oceanic anoxic event 2: Sea surface temperature records from the equatorial Atlantic. Paleoceanography 22, PA1219, doi: 10.1029/2006PA001349.

Frakes L.A. (1999) Estimating the global thermal state from Cretaceous sea surface and continental temperature data. From: Barrera E. & Johnsen C.C. (eds.), Evolution of the Cretaceous Ocean-Climate System, Geological Society of America Special Paper 332, 49-57.

Friedrich O., Nishi H., Pross J., Schmiedl G. & Hemleben C. (2005) Millenial- to Centennial-Scale Interruptions of the Oceanic Anoxic Event 1b (Early Albian, mid-

Cretaceous) Inferred from Benthic Foraminiferal Repopulation Events. Palaios 20, 64-77.

Gérard J.-C. & Dols V. (1990) The warm Cretaceous climate: Role of the long-term carbon cycle. Geophysical Research Letters 17, 1561-1564.

Gradstein F.M., Agterberg F.P., Ogg J.G., Hardenbol J., van Veen P., Thierry J. & Huang Z. (1994) A Mesozoic time scale. Journal of Geophysical Research 99 (B12), 24,051-24,074.

Gwinner M.P. (1971) Geologie der Alpen. Stratigraphie, Paläogeographie, Tektonik.- Schweizerbart Verlag, pp. 477.

Hallam A. & Wignall P.B. (1999) Mass extinctions and sea-level changes. Earth Science Reviews 48 (4), 217-250.

Handoh I.C., Bigg G.R. & Jones E.J.W. (2003) Evolution of upwelling in the Atlantic Ocean basin. Palaeogeography, Palaeoclimatology, Palaeoecology 202, 31–58.

Haq B.U., Hardenbol J. & Vail P.R. (1987) Chronology of Fluctuating Sea Levels Since the Triassic. Science 235, 1156-1167.

Hay W.W., DeConto R.M., Wold C.N., Wilson K.M., Voigt S., Schulz M., Rossby Wold A., Dullo W.-C., Ronov A.B., Balukhovsky A.N. & Söding E. (1999) Alternative global Cretaceous paleogeography. From: Barrera E. & Johnsen C.C. (eds.), Evolution of the Cretaceous Ocean-Climate System, Geological Society of America Special Paper 332, 1-45.

Hay W.W. (2008) Evolvind ideas about the Cretaceous climate and ocean circulation. Cretaceous Research 29, 725-753.

Hedges J.I., Hu F.S., Devol A.H., Hartnett H.E., Tsamakis E. & Keil R.G. (1999) Sedimentary organic matter preservation: a test for selective degradation under oxic conditions. American Journal of Science 299, 529–555.

Herrle J.O. (2002) Paleoceanographic and paleoclimatic implications on mid-Cretaceous black shale formation in the Vocontian Basin and the Atlantic: evidence from calcareous nannofossils and stable isotopes. Tübinger Mikropaläontologische Mitteilungen 27, 1-114.

Herrle J.O. (2003) Reconstructing nutricline dynamics of mid-Cretaceous oceans: evidence from calcareous nannofossils from the Niveau Paquier black shale (SE France). Marine Micropaleontology 903, 1-15.

Herrle J.O., Pross J., Friedrich O. & Hemleben C. (2003) Short-term environmental changes in the Cretaceous Tethyan Ocean: micropalaeontological evidence from the Early Albian Oceanic Anoxic Event 1b. Terra Nova 15, 14-19.

Herrle J.O. & Mutterlose J. (2003) Calcareous nannofossils from the Aptian-Lower Albian of southeast France: palaeoecological and biostratigraphic implications. Cretaceous Research 24, 1-22.

Herrle J.O., Kößler P., Friedrich O., Erlenkeuser H. & Hemleben C. (2004) High-resolution carbon isotope records of the Aptian to Lower Albian from the SE France and the Mazagan Plateau (DSDP Site 545): a stratigraphic tool for paleoceanographic and paleobiologic reconstruction. Earth and Planetary Science Letters 218, 149-161.

Hesselbo S.P., Gröcke D.R., Jenkyns H.C., Bjerrum C.J., Farrimond P., Morgans Bell H.S. & Green O.R. (2000) Massive dissociation of gas hydrate during a Jurassic oceanic anoxic event. Nature 406, 392-395.

Higgins J.A. & Schrag D.P. (2006) Beyond methane: Towards a theory for the Paleocene-Eocene Thermal Maximum. Earth and Planetary Science Letters 245, 523-537.

Hinz K. & Winterer E.L. (Eds.) (1984) Site 545. Shipboard Scientific Party. In: Initial Reports of the Deep Sea Drilling Projects, Vol. 79, 81-177. U.S. Government Printing Office, Washington D.C.

Hofmann P. M. (1992) Sedimentary facies, organic facies, and hydrocarbon generation in evaporite sediments of the Mulhouse Basin, France. Dissertation, Forschungszentrum Jülich, 288 S.

Hofmann P., Ricken W., Schwark L. & Leythaeuser D. (1999) Coupled oceanic effects of climatic cycles from late Albian deep-sea sections of the North Atlantic. From: Barrera E. & Johnsen C.C. (eds.), Evolution of the Cretaceous Ocean-Climate System, Geological Society of America Special Paper 332, 143-159.

Hofmann P., Ricken W., Schwark L. & Leythaeuser D. (2000) Carbon-sufur-iron relationships and $\delta^{13}C$ for organic matter for late Albian sedimentary rocks from the North Atlantic Ocean: Paleoceanographic Implications. Paleogeography, Paleoclimatology, Paleoecology 162, 97-113.

Hofmann P., Ricken W., Schwark L. & Leythaeuser D. (2001) Geochemical signature and related climatic-oceanographic processes for early Albian black shales: Site 417D, North Atlantic Ocean. Cretaceous Research 22, 243-257.

Hofmann P., Stüsser I., Wagner T., Schouten S. & Sinninghe Damsté J.S. (2008) Climate-ocean coupling off North-West Africa during the Lower albian: The Oceanic Anoxic Event 1b. Palaeogeography, Palaeoclimatology, Palaeoecology 262, 157-165.

Holbourn A., Kuhnt W. & Soeding E. (2001) Atlantic paleobathymetry, paleoproductivity and paleocirculation in the late Albian: the benthic foraminiferal record. Palaeogeography, Palaeoclimatology, Palaeoecology 170, 171-196.

Huber B.T., Norris R.D. & MacLeod K.G. (2002) Deep-sea paleotemperature record of extreme warmth during the Cretaceous. Geology 30 (2), 123-126.

Huguet C., Schimmelmann A., Thunell R., Lourens L.J., Sinninghe Damsté J.S. & Schouten S. (2007) A study of the TEX86 paleothermometer in the water column and sediments of the Santa Barbara Basin, California. Paleoceanography 22.

Irving E., North F.K. & Couillard R. (1974) Oil, climate and tectonics. Canadian Journal of Earth Sciences 11, 1-17.

Jahren A.H., Arens N.C., Sarmiento G., Guerrero J. & Amundson R. (2001) Terrestrial record of methane hydrate dissociation in the Early Cretaceous. Geology 29 (2), 159-162.

Jahren A.H. (2002) The biogeochemical consequences of the mid-Cretaceous superplume. Journal of Geodynamics 34, 177-191.

Jenkyns H.C. (1980) Cretaceous anoxic events: from continents to oceans. J. geol. Soc. London 137, 171-188.

Jenkyns H.C. (2003) Evidence for rapid climate change in the Mesozoic-Palaeogene greenhouse world. Philosophical Transactions – Royal Society of London 361, 1885-1916.

Jenkyns H.C., Forster A., Schouten, S. & Sinninghe Damsté J.S. (2004) High temperatures in the Late Cretaceous Arctic Ocean. Nature 432, 888-892.

Kemp D.B., Coe A.L., Cohen A.S. & Schwark L. (2005) Astronmical pacing of methane release in the early Jurrasic period. Nature 437 (15), 396-399.

Kerr R.A. (2001) Paleontology: Paring down the big five mass extinctions. Science 294 (5549), 2072-2073.

Kim J., Schouten S., Hopmans E.C., Donner B. & Sinninghe Damsté J.S. (2008). Global core-top calibration of the TEX86 paleothernometer in the ocean. Geochimica et Cosmochimica Acta 72, 1154–1173.

Kuespert W. (1982) Environmental changes during oil shale deposition as deduced from stable isotope ratios. In: Einsele G. & Seilacher A. (eds.) Cyclic and Event Stratification. Springer-Verlag, Berlin, 482–501.

Kuypers M.M.M., Pancost R.D. & Damsté J.S.S. (1999) A large and abrupt fall in atmospheric CO_2 concentration during Cretaceous times. Nature 399, 342-345.

Kuypers M.M.M., Blokker P., Erbacher J., Kinkel H., Pancost R.D., Schouten S. & Damsté J.S.S. (2001) Massive Expansion of Marine Archaea During a Mid-Cretaceous Oceanic Anoxic Event. Science 293, 92-94.

Kuypers M.M.M., Blokker P., Hopmans E.C., Kinkel H., Pancost R.D., Schouten S. & Sinninghe Damsté J.S. (2002)a Archaeal remains dominate marine organic matter from the early Albian oceanic anoxic event 1b. Palaeogeography, Palaeoclimatology, Palaeoecology 185, 211-234.

Kuypers M.M.M., Nijenhuis I.A. & Sinninghe Damsté J.S. (2002)b Enhanced productivity led to increased organic carbon burial during the late Cenomanian oceanic anoxic event. Paleoceanography 17 (4), 3-1 – 3-13.

Kuypers M.M.M., Lourens L.J., Rijpstra W.R.C., Pancost R.D., Nijenhuis I.A. & Sinninghe Damsté J.S. (2004) Orbital forcing of organic carbon burial in the proto-North Atlantic during oceanic anoxic event 2. Earth and Planetary Science Letters 228 (3–4), 465–482.

Kvenvolden K.A. (1998) A primer on the geological occurrence of gas hydrate. In: Henriet J.-P. & Mienert J. (eds.) Gas Hydrates: Relevance to World Margin Stability and Climatic Change. Geological Society 137. Special Publication, London, 9–30.

Kvenvolden K.A. & Lorenson T.D. (2001) Attention turns to naturally occurring methane seepage. EOS Trans., AGU 82 (40), 457.

Langford F.F. & Blanc-Valleron M.M. (1990) Interpreting Rock-Eval pyrolysis data using graphs of pyrolizable hydrocarbons vs. total organic carbon. AAPG Bulletin 74 (6), 799-804.

Larson R.L. (1991) Geological consequences of superplumes. Geology 19, 963-966.

Larson R.L. & Erba E. (1999) Onset of the mid-Cretaceous greenhouse in the Barremian-Aptian: Igneous events and the biological, sedimentary, and geochemical responses. Paleoceanography 14 (6), 663-678.

Leckie R.M. (1984) Mid-Cretaceous planktonic foraminiferal biostratigraphy off central Morocco, Deep Sea Drilling Project Leg 79, Site 545 and 547. Initial Report Deep Sea Drilling Project 79, 579-620.

Leckie R.M., Bralower T.J. & Cashman R. (2002) Oceanic anoxic events and plankton evolution: Biotic response to tectonic forcing during the mid-Cretaceous. Paleoceanography 17 (3), 13-1 – 13-29.

Lemoine M. (1984) Mesozoic evolution of the western Alps. Annales Geophysicae 2 (2), 171-172.

Lemoine M. (1985) Structuration jurassique des Alpes occidentales et palinspastiques de la Téthys ligure. Bulletin de la Société Géologique de France 8 (1), 126-137.

Lever A. & McCave N. (1983) Eolian components in Cretaceous and Tertiary North Atlantic sediments. Journal of Sedimentary Research 53 (3), 811-832.

Levitus S., Boyer T., Burgett R. & Conkright M. (1998) World ocean atlas 1998 (WOA98): Ocean Climate Laboratory, National Oceanographic Data Center (data accessible at http:// ferret.wrc.noaa.gov/las/main.html).

Lorenz C. (1980) Géologie des pays européens, France, Belgique, Luxembourg.- Paris.

Martinez P., Bertrand P., Shimmield G.B., Cochrane K., Jorissen F.J., Foster J. & Dignan M. (1999) Upwelling intensity and ocean productivity changes of Cape Blanc (northwest Africa during the last 70 000 years: geochemical and micropalaeontological evidence. Marine Geology 158, 57–74.

Meehl G.A., Stocker T.F., Collins W.D., Friedlingstein P., Gaye A.T., Gregory J.M., Kitoh A., Knutti R., Murphy J.M., Noda A., Raper S.C.B., Watterson I.G., Weaver A.J. & Zhao Z.- C. (2007) Global Climate Projections. In: Climate Change 2007: The Physical Science Basis. Contribution of Working Group I to the Fourth Assessment Report of the Intergovernmental Panel on Climate Change. Solomon S., Qin D., Manning M., Chen Z., Marquis M., Averyt K.B., Tignor M. & Miller H.L. (eds.). Cambridge University Press, Cambridge, United Kingdom and New York, NY, USA.

Mhammdi N. & Bobier C. (2001) The sedimentary system tracts on the Moroccan Atlantic margin. Earth system processes – Global meeting (June 24-28, 2001). Conference Abstract

Milkov A.V. (2004) Global estimates of hydrate-bound gas in marine sediments: how much is really out there? Earth-Sci. Rev. 66 (3–4), 183–197.

Moreno A., Taragrona J., Hendriks J., Canals M., Freudenthal T. & Meggers H. (2001) Orbital forcing of dust supply to the North Canary Basin over the past 250 kyrs. Quaternary Science Reviews 20, 1327–1339.

Moullade M. (1966) Etude stratigraphique et micropaléontologique de Crétacé inférieur de la « fosse vocontienne ». Documents de la Laboratoire de Géologie de la Faculté de Sciences, Lyon 15, 1-369.

Nesbitt H.W. & Young G.M. (1982) Early Proterozoic climates and plate motions inferred from major element chemistry of lutites. Nature 299, 715-717.

Nesbitt H.W. & Young G.M. (1996) Petrogenesis of sediments in the absence of chemical weathering: effects of abrasion and sorting on bulk composition and mineralogy. Sedimentology 43, 341-358.

Norris R.D. & Roehl U. (1999) Carbon cycling and chronology of climate warming during the Palaeocene/Eocene transition. Nature 401, 775–778.

Norris D. R., Bice K. L., Mango E. A. & Wilson P. A. (2002) Jiggling the tropical thermostat in the Cretaceous hothouse. Geology 30, 299– 302.

Ogg J.G., Agterberg F.P. & Gradstein F.M. (2004) The Cretaceous Period. From: Gradstein F.M., Ogg J.G. & Smith A.G. (eds) A geologic time scale 2004, Cambridge University Press, 344-383.

Olivero D. & Mattioli E. (2008) The Aalenian-Bajocian (Middle Jurrasic) of the Digne area. In: Mattioli M., Gardin S., Giraud F., Olivero D., Pittet B. & Reboulet S. (eds) Guidebook for the post-congress fieldtrip in the Vocontian Basin, SE France (September 11-13, 2008). Carnets de Géologie / Notebooks on Geology, 30 S.

Peters K.E. (1986) Guideline for evaluating petroleum source rock using programmed pyrolysis. The American Association of Petroleum Geologists Bulletin 70, 318-396.

Peters K.E., Walters C.C. & Moldowan J.M. (2005) The biomarker guide: Volume 2. Cambridge University Press, 471 pp., -2nd Edition. United Kingdom.

Pittet B. (2008) The Cenozoic of the Barrême syncline. In: Mattioli M., Gardin S., Giraud F., Olivero D., Pittet B. & Reboulet S. (eds) Guidebook for the post-congress fieldtrip in the Vocontian Basin, SE France (September 11-13, 2008). Carnets de Géologie / Notebooks on Geology, 30 S.

Pletsch T., Daoudi L., Chamley H., Deconinck J.F. & Charroud M. (1996) Palaeogeographic controls on palygorskite occurrence in mid-Cretaceous sediments of Morocco and Adjacent Basins. Clay Minerals 31, 403-416.

Pletsch T. (1998) Origin of Lower Eocene Palygorskite clays on the Cote d´Ivoire-Ghana transform margin, Easrten equatorial Atlantic. From: Mascle J., Lohmann G.P & Moullade, M. (eds) Proceedings of the Ocean Drilling Program, Scientific Results 159, 141-156.

Pletsch T., Erbacher J., Holbourn A.E.L., Kuhnt W., Moullade M., Oboh-Ikuenobe F.E., Söding E. & Wagner T. (2001): Cretaceous separation of Africa and South America: the view from the West African margin (ODP Leg 159).– J. South Amer. Earth Sci. 14, 147–174.

Pletsch T. (2003) Palaeoenvironmental implications of palygorskite clays in Eocene deep-water sediments from the western Central Atlantic. Geological Society, London, Special Publications, 2001; 183, 307-316.

Popp B., Takigiku R., Hayes J.M., Louda J.W. & Baker E.W. (1989) The post Paleozoic chronology and mechanism of 13C depletion in primary organic matter. Am. J. Sci. 289, 436–454.

Poulsen C.J., Seidov D., Barron E.J. & Peterson W.H. (1998) The impact of palaeogeographic evolution of the surface oceanic circulation and the marine environment within the mid-Cretaceous Tethys. Palaeoceanography, 13, 546-559.

References

Poulsen C.J., Barron E.J., Johnson C.C. & Fawcett P. (1999) Links between major climatic factors and regional oceanic circulation in the mid-Cretaceous. From: Barrera E. & Johnsen C.C. (eds.), Evolution of the Cretaceous Ocean-Climate System, Geological Society of America Special Paper 332, 91-103.

Poulsen C. L., Barron E. J., Arthur M. A. & Peterson W. H. (2001) Response of mid-Cretaceous global oceanic circulation to tectonic and CO_2 forcings. Paleoceanography 16, 576–592.

Premoli Silva I. & Sliter W.V. (1999) Cretaceous paleoceanography: Evidence from planktonic foraminiferal evolution. From: Barrera E. & Johnsen C.C. (eds.), Evolution of the Cretaceous Ocean-Climate System, Geological Society of America Special Paper 332, 301-328.

Raiswell R. & Berner R.A. (1985) Pyrite formation in euxinic and semieuxinic sediments. American Journal of Science 285, 710-724.

Sageman B.B. & Hollander D.J. (1999) Cross correlation of paleoecological and geochemical proxies: A holistic approach to the study of past global change. From: Barrera E. & Johnsen C.C. (eds.), Evolution of the Cretaceous Ocean-Climate System, Geological Society of America Special Paper 332, 365-384.

Savostin L.A., Sibuet J., Zonenshain L.P., LePichon X. & Poulet M.J. (1986) Kinematic evolution of the Tethys belt from Atlantic Ocean to Pamirs since Triassic. Tectonophysics, 123, 1-35.

Schlanger S.O. & Jenkyns H.C. (1976) Cretaceous oceanic anoxic events: causes and consequences. Geologie en Mijnbouw 55 (3-4), 179-184.

Schettino A. & Scotese C. (2002) Global kinematic constraints to the tectonic history of the Mediterranean region and surrounding areas during the Jurrasic and Cretaceous. In: Rosenbaum G. & Lister G.S. (2002) Reconstruction of the evolution of the Alpine-Himalayan Orogen. Journal of the Virtual Explorer 8, 145-160.

Schouten S., Rjipstra W.I.C., Sinninghe Damsté J.S., de Leeuw J.W. & Schoell M. (1997a) A molecular stable carbon isotope study of organic matter in immature Miocene Monterey sediments, Pismo Basin. Geochimica et Cosmochimica Acta 61, 2065-2082.

Schouten S., Hoefs M. J. L. & Sinninghe Damsté J. S. (2000) A molecular and stable carbon isotopic study of lipids in late Quaternary sediments from the Arabian Sea. Org. Geochem., 31, 509–521.

Schouten S., Hopmans E.C., Schefuß E. & Sinninghe Damsté J.S. (2002) Distributional variations in marine crenarchaeotal membrane lipids: a new tool for reconstructing ancient sea water temperatures? Earth and Planetary Science Letters 204, 265-274.

Schouten S., Hopmans E.C., Forster A., van Breugel Y., Kuypers M.M.M. & Sinninghe Damsté J.S. (2003a) Extremely high sea-surface temperatures at low latitudes during the middle Cretaceous as revealed by archaeal membrane lipids. Geology 31 (12), 1069-1072.

Schouten S., Hopmans E.C., Schefuß E. & Sinninghe Damsté J.S. (2003b) Distributional variations in marine crenarchaeotal membrane lipids: a new tool for reconstructing ancient sea water temperatures? Corrigendum: Earth and Planetary Science Letters 211, 205-206.

Schouten S., Rampen S.W., Geenevasen J.A.J. & Sinninghé Damste J.S. (2005) Structural identification of steryl alkyl ethers in marine sediments. Org.Geochem., 36, 1323-1333.

Schouten S., Huguet C., Hopmans E.C. & Sinninghe Damsté J.S. (2007a) Improved analytical methodology of the TEX86 paleothermometry by high performance liquid chromatography/atmospheric pressure chemical ionization-mass spectrometry. Analytical Chemistry 79, 2940–2944.

Scotese C.R. (1998) Plate Tectonics. In: Encyclopedia of the Earth Sciences, Macmillan Publishing Company, New York, 859-864.

Scotese C.R. (2001) Atlas of Earth History, PALEOMAP Project. Arlington, Texas, 52 pp.

Shimmield G.B. (1992) Can sediment geochemistry record changes in coastal upwelling paleo-productivity? Evidence from northwest Africa and the Arabian Sea. In: Summerhayes C.P., Prell W.E. & Emeis K.C. (eds.) Upwelling Systems: Evolution Since the Early Miocene. Geological Society Special Publication 64. The Geological Society, London, 29–46.

Sinninghe Damsté J. S., Kuypers M. M. M., Schouten S., Schulte S. & Rullkötter J. (2003) The lycopane/C31 n-alkane ratio as a proxy to assess palaeooxicity during sediment deposition, Earth Planet. Sci. Lett., 209, 215–226.

Skelton P.W (ed.) (2003) The Cretaceous World. Cambridge University Press, 360 pp., United Kingdom.

Sprovieri M., Coccioni R., Lirer F., Pelosi N. & Lozar F. (2006) Orbital tuning of lower Cretaceous composite record (Maiolica Formation, central Italy). Paleoceanography 21, PA4212. doi: 10.1029/2005PA001224.

Stein R. & Littke R. (1990) Organic carbon rich sediments and paleoenvironment: Results from Baffin Bay (ODP Leg 105) and the upwelling area off northwest Africa (ODP Leg 108). In: Huc A.Y. (ed) Deposition of Organic Facies. AAPG Studies in Geology 30, 41-56. American Association of Petroleum Geologists, Tulsa.

Steuber T., Rauch M., Masse J.-P., Graaf J. & Malkoc M. (2005) Low-latitude seasonality of Cretaceous temperatures in warm and cold episodes. Nature 437, 1341-1344.

Takashima R., Nishi H., Huber B. T. and Leckie R. M. (2006) Greenhouse world and the Mesozoic ocean. Oceanography 19, 64-74.

Taylor S.K. & McLennan S.M. (1985) The Continental Crust, Its Composition and Evolution. Blackwell Science, Oxford.

Thomas D.J., Zachos J.C., Bralower T.J., Thomas E. & Bohaty S. (2002) Warming the fuel for the fire: Evidence for the thermal dissociation of methane hydrate during the Paleocene-Eocene thermal maximum. Geology 30 (12), 1067-1070.

Tribovillard N., Algeo T.J., Lyons T. & Riboulleau A. (2006) Trace metals as paleoredox and productivity proxies: an update. Chemical Geology 232, 12–32.

Tsikos H., Karakitsios V., van Breugel Y., Walsworth-Bell B., Bombardiere L., Petrizzo M.R., Sinninghe Damsté J.S., Schouten S., Erba E., Premoli Silva I., Farrimond P., Tyson R.V. & Jenkyns H.C. (2004) Organic-carbon deposition in the Cretaceous of the Ionian Basin, NW Greece: the Paquier Event (OAE 1b) revisited. Geological Magazine 141 (4), 401-416.

Twitchett R.J. (2006) The palaeoclimatology, palaeoecology and palaeoenvironmental analysis of mass extinction events. Palaeogeography, Palaeoclimatology, Palaeoecology 232 (2-4), 190-213.

Tyson R.V. (1995) Sedimentary organic matter: Organic facies and palynofacies. Chapman & Hall. 615 pp., London, UK. (Pages 22-47, 118-149, 367-383, 394-417.)

Ufnar D.F., González L.A., Ludvigson G.A., Brenner R.L. & Witzke B.J. (2004). Evidence for increased latent heat transport during the Cretaceous (Albian) greenhouse warming. Geology 32, 1049–1052.

Vink A. (1996) Organic geochemical characterization of the Lower Albian Niveau Paquier; A key black shale horizon from the Vocontian Basin (SE France) 52 pp. unpublished.

Vink A., Schouten S., Sephton S. & Sinninghe Damsté J.S. (1998) A newly discovered norisoprenoid, 2,6,15,19-tetramethylicosane, in Cretaceous black shales. Geochimica et Cosmochimica Acta 62 (6), 965-970.

Voigt S. (1996) Paläobiogeographie oberkretazischer Inoceramen und Rudisten.- Münchner Geowissenschaftliche Abhandlungen, 31, 1-101.

Voigt S., Hay W.H., Höfling R. & DeConto R.M. (1999) Biogeographic distribution of Late Cretaceous rudist-reefs in the Mediterranean as climate indicators. From: Barrera E. & Johnsen C.C. (eds.), Evolution of the Cretaceous Ocean-Climate System, Geological Society of America Special Paper 332, 91-103.

Voigt S., Gale A.S. & Flögel S. (2004) Midlatitude shelf seas in the Cenomanian-Turonian greenhouse world: Temperature evolution and North Atlantic circulation. Paleoceanography 19 (4), 1-17.

Wagner T. & Pletsch T. (1999) Tectono-sedimentary controls on Cretaceous black shale deposition along the opening Equatorial Atlantic Gateway (ODP Leg 159). From: Cameron N., Bate R. & Clure V. (eds) The Oil and Gas Habitat of the South Atlantic, Geological Society Special Publication.

Wagner T., Sinninghe Damsté J.S., Hofmann P. & Beckmann B. (2004) Euxinia and primary production in Late Cretaceous eastern equatorial Atlantic surface waters fosterd orbitally driven formation of marine black shales. Paleoceanography 19, PA3009, doi:10.1029/2003PA000898.

Wagner T., Wallmann K., Herrle J.O., Hofmann P. & Stuesser I. (2007) Consequences of moderate ~25,000 yr lasting emission of light CO_2 into the mid-Cretaceous ocean. Earth and Planetary Science Letters 259, 200-211.

Wagner T., Herrle J.O., Sinninghe Damsté J.S., Schouten S., Stüsser I. & Hofmann P. (2008) Rapid warming and salinity changes of Cretaceous surface waters in the subtropical North Atlantic. Geology 36 (3), 203-206.

Weaver C.E. & Pollard L.D. (1975) The chemistry of clay minerals. Developments in Sedimentology, 15. Elsevier, Amsterdam.

Weaver C.E. (1989) Clays, muds, and shales. Developments in Sedimentology, 44. Elsevier, Amsterdam.

Weissert H. & Bréhéret J.G. (1991) A carbonate-carbon isotope record from Aptian -Albian sediments of the Vocontian Trough. Bull. Soc. Géol. Fr. 162, 1133-1140.

Weissert H., Lini A., Föllmi K.B. & Kuhn O. (1998) Correlation of Early Cretaceous carbon isotope stratigraphy and platform drowning events: a possible link? Palaeogeography, Palaeoclimatology, Palaeoecology 137, 189-203.

Wilpshaar M. & Leereveld H. (1994) Palaeoenvironmental change in the Early Cretaceous Vocontian Basin (SE France) reflected by dinoflagellate cysts. Review of Palaeobotany and Palynology, 84, 121-128.

Wilpshaar M., Leereveld H. & Visscher H. (1997) Early Cretaceous sedimentary and tectonic development of the Dauphinois Basin (SE-France). Cretaceous Research, 18, 457-468.

Wilson P.A. & Norris R.D. (2001) Warm tropical ocean surface and global anoxia during the mid-Cretaceous period. Nature 412, 425-429.

REFERENCES

Wilson P.A., Norris R.D. & Cooper M. J. (2002) Testing the Cretaceous greenhouse hypothesis using glassy foraminiferal calcite from the core of the Turonian tropics on Demerara Rise. Geology 30 (7), 607-610.

Zabel M., Schneider R.R., Wagner T., Adegbie A.T., de Vries U. & Kolonic, S. (2001) Late Quaternary climate changes in Central Africa as inferred from terrigenous input to the Niger Fan. Quaternary Research 56, 207–217.

Online:

Fig. 5: picture available under: http://jan.ucc.nau.edu/~rcb7/midcretmed.jpg

Fig. 7: picture available under:
http://www.usgcrp.gov/usgcrp/Library/nationalassessment/LargerImages/SectorGraphics/Coastal/Belt.jpg

Fig. 14: map created with OMC 2004 by M. Weinelt, available under:
http://www.aquarius.geomar.de/omc/omc_intro.html.

Die VDM Verlagsservicegesellschaft sucht für wissenschaftliche Verlage abgeschlossene und herausragende

Dissertationen, Habilitationen, Diplomarbeiten, Master Theses, Magisterarbeiten usw.

für die kostenlose Publikation als Fachbuch.

Sie verfügen über eine Arbeit, die hohen inhaltlichen und formalen Ansprüchen genügt, und haben Interesse an einer honorarvergüteten Publikation?

Dann senden Sie bitte erste Informationen über sich und Ihre Arbeit per Email an *info@vdm-vsg.de*.

Sie erhalten kurzfristig unser Feedback!

VDM Verlagsservicegesellschaft mbH
Dudweiler Landstr. 99   Telefon +49 681 3720 174
D - 66123 Saarbrücken   Fax   +49 681 3720 1749
www.vdm-vsg.de

Die VDM Verlagsservicegesellschaft mbH vertritt

Printed by Books on Demand GmbH, Norderstedt / Germany